ANCIENT PIONEERS

THE FIRST AMERICANS

Trio of 500-year-old hollow silver figurines, Inka culture, Peru

NATIONAL
GEOGRAPHIC

WASHINGTON, D.C.

ANCIENT PIONEERS
THE FIRST AMERICANS

BY GEORGE E. STUART

CONTENTS

INTRODUCTION   6

CHAPTER 1   BEGINNINGS   12

CHAPTER 2   THE POLAR REGIONS   28

CHAPTER 3   THE WEST   48

CHAPTER 4   THE SOUTHWEST   68

CHAPTER 5   THE SOUTHEAST   92

CHAPTER 6   MESOAMERICA   116

CHAPTER 7   THE ANDES & BEYOND   148

EPILOGUE   190

AUTHOR NOTES, ACKNOWLEDGMENTS,   194
ILLUSTRATION CREDITS

INDEX   195

Polar bears stalk their frozen Arctic home in search of food as did many of the earliest American hunters who entered the Americas by way of the far north.

PRECEDING PAGES: *The brooding mass of the Pyramid of the Sun at Teotihuacan, Mexico, was built to represent its natural counterpart—the mountain. Such structures stand as the ultimate overt acknowledgment of land, landscape, and nature in the mythology and history of many ancient American cultures.*

Aleutian Is.
Anangula Island
Bering Sea
Siberia
Bering Str.
Chukchi Sea
Ipiutak
Arctic Ocean
GREENLAND
Onion Portage
Noatak
Kobuk
Alaska
Kongakut R.
Gulf of Alaska
Baffin Bay
Qilakitsoq
Baffin I.
Baffin
Davis Strait
Sitka
Whitefish L.
Labrador Sea
Masset
Skidegate
Kitwanga
Queen Charlotte Is.
ROCKY MTS.
Great Plains
Canadian Shield
Hudson Bay
Labrador
Newfoundland
Vancouver I.
Cape Flattery
Ozette
Fort Rock Cave
Wilson Butte Cave
Lassen Peak
Danger Cave
San Francisco Bay
Great Basin Desert
Tuolumne Pass, Yosemite
Monument Valley
Grand Canyon
Cuyahoga R.
Mts.
Shawnee-Minisink
Santa Catalina
Ship Rock
Koster
Chillicothe
24 25 26
Meadowcroft
Folsom
Cahokia
Ohio
27
23
Mississippi
Appalachian
Cactus Hill
Salt
Clovis
Spiro
Eva
Secotan
Gila
Lewisville
Pinson
Mulberry site/Cofitachequi
San Pedro R.
Levi
Stallings Island
Marksville
Savannah R.
Weeden I.
Fort Center
Gulf of Mexico
Bahamas
Atlantic Ocean
1. Keet Seel
2. Betatakin
3. Long House Valley
4. Canyon de Chelly
5. Cowboy Wash
6. Mesa Verde
7. Chaco Canyon
8. El Morro Valley
9. Pine Lawn Valley
10. Mimbres Valley
11. Paquimé (Casas Grandes)
12. Ventana Cave
13. Casa Grande
14. Snaketown

15. Watson Brake
16. Poverty Point
17. Moundville
18. Mandeville
19. Kolomoki/Swift Creek
20. Ocmulgee
21. Etowah
22. Ayers Mound
23. Serpent Mound
24. Newark
25. Flint Ridge
26. Grave Creek
27. Cresap

Teotihuacán
Cacaxtla
Oxkintok
Uxmal
Cuba
Tenochtitlan
Tula
Chichén Itzá
La Isabela
Puerto Rico
Virgin Is.
Xochicalco
Calakmul
Hispaniola
Greater Antilles
Chalcatzingo
28 29 30
Rio Azul
Teopantecuanitlán
31
Uaxactún
Tlapacoya/Valsequillo
Monte Albán
San Lorenzo
Tikal
El Manatí
Palenque
Quiriguá
Caribbean Sea
Lesser Antilles
Piedras Negras
Copán
Bonampak
Kaminaljuyú
Lake Managua
Pueblito
Puerto Hormiga
28. Cerro de las Mesas
29. Tres Zapotes
30. Laguna de los Cerros
31. La Venta
Rivas
Guayabo de Turrialba
Nicoya Pen.
Sitio Conte
Coclé
Pacific Ocean
Galápagos Is.
San Agustín
Caverna de Pedra Pintada
Negro
AMAZON
Marajó I.
Valdivia
Amazon
Santarém
BASIN
Pampa Grande
Sipán
ANDES
Cupisnique
Cajamarca
Huaca Prieta
Huaca del Sol
Chan Chan
Kotosh
Brazilian Highlands
Chavín de Huantar
Pikimachay Cave
Aspero
Wari
El Paraíso
Machu Picchu
La Florida
Cuzco (Cusco)
Mina Perdida
Paracas
Lake Titicaca
Cahuachi
Tiwanacu
Cerro Llullaillaco
Quebrada las Conchas
Gran Chaco
ANDES
Atacama Desert
Pampas
Maule R.
Patagonia
Monte Verde
Tierra del Fuego

ANCIENT PIONEERS

■ Place mentioned in text
— Present-day country boundary

0   1000 miles
0   1000 kilometers

Bipolar Oblique projection

INTRODUCTION

Together, the Americas cover one-quarter of the Earth's habitable surface and hold virtually every major climatic zone that exists on the planet. The peopling of these varied landscapes by the distant ancestors of the present-day Native Americans— the Indians and Eskimos of history—took place somewhere between 12,000 and 20,000 years ago. By the end of the 15th century A.D., tens of millions of people speaking some 2,000 mutually unintelligible languages occupied nearly every livable part of the Western Hemisphere, with each group representing the culmination of a long and complicated cultural path. This map depicts the major sites of ancient camping places, settlements, villages, and cities. Each played a part in the long and complicated story that ended with the appearance of the conquerors and colonists who followed in the wake of the European "discovery."

THE TOWN LAY BESIDE A SHALLOW RIVER of medium width, on a flat, slightly elevated plain by the mouth of a creek. Both the river and the creek were somewhat higher and faster than usual, for spring rains had done their good work and cornfields would soon burst forth from the fertile alluvial soil. The town itself consisted of a large cluster of perhaps a hundred sturdy dwellings whose plastered clay walls seemed part of the land from which they rose. These houses were neatly roofed with split-cane mats, mostly browned with age, though some newly finished roofs shone green in the morning brightness.

Three buildings—one in particular—dominated the town, for they overlooked it from flattopped mounds of earth. The mound beside the river rose much higher than the others. It held the largest building, one whose doorway faced east, away from the river, toward the rising sun and a large open plaza surfaced with smooth clay. The colors made the whole settlement seem part of the land itself, except for the signs and sounds of human presence: the acrid pall of wood smoke, the barking of dogs, laughter, an occasional shout, and the murmur of conversation that carried well in the cool air.

On the morning of May 1, 1540, according to the Christian calendar that they observed, a small group of men carefully watched the town from the opposite bank. They were tired, for they had been walking trails and crossing rivers and trackless swamps of a hot, humid, and unknown land for almost a year. They were Spaniards or, more accurately, soldiers of the Christian kingdoms of Castile and Aragon—"Spain" would not emerge as a single nation until later—and they considered themselves, as historian Charles Hudson points out, "a specially favored people who were carrying out a divine mission." In practical terms they had come in search of gold and riches like those discovered by their colleagues in the southern lands of Mexico and Peru.

The leader of the small army of entrepreneurs and adventurers was 40-year-old Hernando de Soto, already a veteran of

more than two decades of cruel conquest in Panama, Nicaragua, and Peru. On this campaign he had relentlessly led some 400 members of the expedition, along with more than 200 horses, a herd of pigs, and an assortment of dogs all the way from present-day Tampa Bay, Florida, into central South Carolina. De Soto and an advance guard had reached the Wateree River. The town opposite was Cofitachequi, and it was one of the largest towns the Spaniards had seen—clearly an important place.

The townspeople quickly realized that the strangers they

Cofitachequi town, reconstructed for a diorama in the South Carolina State Museum in Columbia, shows the main buildings on

platform mounds, the great plaza, and surrounding dwellings as they may have looked in the spring of A.D. 1540.

had heard rumor of had arrived. Men soon appeared on the bank opposite the Spaniards, according to Hudson's reconstruction of the occasion, and they carried a litter covered with white cloth and occupied by a woman, clearly one of high rank, whom the Spaniards knew simply as "the Lady of Cofitachequi." Her bearers conveyed her to the base of the trail-worn river-bank, to an orderly row of long, carefully crafted wooden canoes beached there.

"The lady entered a dugout canoe that had an awning over

its stern," continues Hudson. "Beneath the awning a mat was placed on the floor of the canoe, and two cushions were placed on that. She seated herself on the cushions, and several of her principal men climbed aboard. Others climbed into several other canoes and they paddled over to the bank where de Soto stood. The lady got out of the canoe and seated herself on a seat that her subjects had brought for her."

With the help of de Soto's interpreter, the Lady of Cofitachequi welcomed the strangers. She presented them with gifts of animal skins and cloth, then removed several strands of large freshwater pearls from her neck and placed them about de Soto's. Later, all crossed the river to the town, where they visited the great mortuary house on the mound. The interior of the sacred building was filled with the honored dead of Cofitachequi's nobility, each resting in a wooden box and accompanied by many pearls and objects of copper and mica—not gold and silver, as the greedy, disappointed visitors had hoped.

A few days later the remainder of de Soto's huge group arrived. Quickly, and not surprisingly, de Soto's men wore out their welcome by looting the royal corpses and taking hostage the Lady of Cofitachequi (she would soon escape) as they moved on in search of wealth somewhere on the trails ahead.

Cofitachequi survived de Soto's visit, but not for long. According to later descriptions, its population had already begun to decrease severely from a variety of causes ranging from agricultural failure to the onslaught of disease introduced by earlier European explorers.

On the day that the Lady of Cofitachequi cautiously welcomed de Soto to her kingdom, less than half a century had passed since the voyages of Christopher Columbus had "discovered" what Western Europeans regarded as a "New World," ripe for the taking. A scant 21 years before the encounter beside the Wateree River, Hernán Cortés and his men had stood on the chilly barren slope of a snow-covered volcano, regarding with awe the splendid island capital of the Aztec glimmering in the morning sun. Nine months before the Lady of Cofitachequi stepped out of her canoe, Spanish cleric Marcos de Niza had reported to the viceroy in Mexico of the wondrous wealth he had seen among the cities of Cibola, the Zuni towns of the North American Southwest. Each of these encounters between Europeans and those they called "Indians," who had descended from the original inhabitants of the Americas, symbolized the end of an era, for at the time of contact the lives of the innumerable groups of Indians and Eskimos had proceeded relatively uninterrupted for spans of time mostly counted in thousands of years.

We now know with near certainty that the Indians whom de Soto encountered on that particular day in May some 460 years ago occupied what is now known as the Mulberry archaeological site, near Camden, South Carolina. We know that the village the Spaniards walked into was inhabited by speakers of a language ancestral to that of their modern Creek descendants, and that they were part of a larger cultural picture of regional political units headed by a paramount chief—in the case of Cofitachequi, a female leader. We also see a greater picture of the town as part of a highly complex culture that occupied southeastern North America (archaeologists now call it Mississippian) made up of talented corn farmers, skilled artists and craftspeople, and an elite nobility that included shaman-chiefs and war leaders.

The first Americans treated in the chapters that follow were the true pioneers of the continents they came to so very long ago—peoples all but lost in the mists of deep time. Their story is mainly one of successful human adaptation to incredibly varied—and ever changing—environments that stretch from the seemingly inhospitable Arctic plain to the trackless desert country of southern South America. It is a story of great successes achieved by hunters, gatherers, farmers, and householders; by shamans, citybuilders, nobles, kings, and queens. It is an epic of accomplishment in the arts and mathematics, in warfare and trade, and in the building of states and empires. Most of all, it is a story of everyday human beings organized in different ways in countless complex societies—human beings, mostly anonymous now, who more often than not followed with reverence the example of their ancestors and held staunchly to their beliefs and their cultures as an integral part of a cosmos that was at once awesome, sacred, and ever unpredictable.

CHAPTER 1

BEGINNINGS

PIONEER HUNTERS POPULATE
THE AMERICAN CONTINENTS AND
SUCCESSFULLY ADAPT TO
THEIR VARIED AND CHANGING
ENVIRONMENTS

AROUND 18,000 YEARS AGO, during the last of the ice ages, periods of cold that define the Late Pleistocene period, much of Earth's northern waters had frozen into ice sheets and glaciers that covered most of southeastern Alaska and Canada. As a result, the level of the northern seas dropped some 250 to 300 feet, exposing the continental shelves and creating continuous land, virtually a whole subcontinent, connecting Siberia and Alaska. Such a land bridge lay exposed from time to time, with intervening warmer periods that not only re-submerged the land, but also opened ice-free routes to the south. This extreme northwest part of the Americas, most archaeologists agree, served as the main, if not the only, gateway for the true pioneers of North and South America—the hunters, fishers, and gatherers of the period before about 10,000 years ago, whom archaeologists collectively call the Paleo-Indians.

Who were these first human beings to enter the lands we now call America? No question is more fundamental to any consideration of the archaeology of the Western Hemisphere, yet no question is more vexing to archaeologists and others who seek its answer.

Until 1927 archaeologists generally held the view that the first Americans arrived around 5,000 years ago. All that changed after George McJunkin, a cowhand, reported the discovery of some bones exposed in Wild Horse Gulch, near the town of Folsom, New Mexico. The bones turned out to be those of an extinct, long-horned bison of the sort that roamed the Southwest during the last ice age. And, wonder of wonders, the bones at the site of McJunkin's chance discovery lay in direct association with small, carefully made stone weapon points.

"Folsom points," as they came to be called, are distinguished by a long "flute," or wide shallow groove, that occupies all but the very edges of the point on each side. They later proved to date from between 10,000 and 11,000 years ago. The Folsom find, however, was only the beginning of the chase for the earliest Americans: At one of many Folsom sites discovered since, another type of point appeared below—and thus earlier than—the layer containing Folsom points. These distinctive specimens, called Clovis points for the town in New Mexico near where they were first discovered, are usually between three and five inches long, larger than Folsom points, but with a shorter flute on each side, near the base. The discovery of Clovis points pushed the peopling of the Americas back still further, to at least 12,000 years ago—and new research on ice cores may push the date for the earliest Clovis migrants even further into the last ice age.

Like all great works of art and all perfect tools, a Clovis point is a wondrous thing to behold. Each is thin and beautifully flaked, with razor-sharp edges, most often with flutes that made them easier to secure to a wooden shaft. Fluted points are not easy to make, as the late legendary Don Crabtree found after decades in pursuit of success. Crabtree's method, involving a sturdy T-shaped chest press of wood with a piece of antler anchored firmly in the end, was demon-

LEFT: *Gigantic Clovis points, discovered by chance in an orchard near East Wenachee, Washington, indicate the skill of the flint knapper who produced them more than 12,000 years ago. These grand examples of translucent chalcedony may have been for ritual use or were perhaps the remains of a hunter's or knapper's special stash of points. Patterns of flaking on such Paleo-Indian weapon points, along with the relative chronology of Clovis finds throughout North America, have led to a thought-provoking hypothesis by Smithsonian Institution archaeologist Dennis Stanford that at least some Paleo-Indian hunters entered the Americas by means of a northern Atlantic route.*
OPPOSITE: *Caribou, such as this one by the Thelon River in Canada's Northwest Territories, provided food for Paleo-Indian hunters of the far north. Other megafauna in the ancient Americas included mammoths, mastodons, and a now extinct form of bison.*

strated many years ago by another expert knapper, Bruce Bradley.

Much easier said than done, the process required meticulous preparation, then a sudden application of great force by the chest press. First, Bruce secured the prepared point—finished except for the flutes—point down, in a secure crevice on the ground.

Then, with the T-end of the chest press firmly against his body, and the other end against a tiny protrusion he had intentionally left on the base of the point, he lurched suddenly. After that quick motion, honed to success by countless failed practice tries over the years, Bruce noted the result with satisfaction: A flake had been detached from the base of the point and lay by it. Once more on the other face, and he had made a perfect Folsom point. "I've tried this many times," he told me, "Sometimes it works; sometimes it doesn't, but this is how they had to have made these things."

Bradley and his colleagues Dennis Stanford and George Frison have done extensive experimentation in the actual use of Paleo-Indian tools and weapons. Using Clovis points he had made, Bradley secured them with pine pitch and beeswax to wooden shafts. Then, using atlatls, or spearthrowers, to propel these "darts" with great force, or using the tipped shafts as thrusting spears, Bradley, Stanford, and Frison tried them on dead livestock and circus elephants in order to test the penetrating power of the points. They also tested the use of other Clovis blades as butchering tools. Their conclusions: "Clovis weaponry would have been effective for killing mammoths, bison, and other large animals."

Remains of Clovis hunters have come to light at camp and kill sites from Alberta to Panama, and from eastern Canada south to Florida. Those that can be dated securely and consistently date from between about 11,000 and 12,000 years ago. Others, such as surface finds on the frozen tundra—impossible to date—

More than 12,000 years ago, in what is now southern Chile, an ice age hunter in pursuit of a group of llama-like prey brandishes a bola and a sharpened wooden lance. Such weapons were among a unique array of dated

artifacts found by University of Kentucky archaeologist Thomas Dillehay and his Chilean colleagues at Monte
Verde. Their work has helped prove that people first entered the Americas much earlier than previously thought.

nonetheless help to indicate the wide range of what must have been the single most successful hunting weapon in the cultural history of the Western Hemisphere.

Clovis and Folsom hunters and their Paleo-Indian contemporaries throughout the Americas sought and killed large animals—the woolly mammoth, the mastodon, the long-horned bison—the likes of which we will never see in the wild, for they became extinct as warmer climates ended the Ice Age some 10,000 years ago. But the large animals were not the only quarry. At the Levi rockshelter in central Texas, archaeologists found no evidence of large mammals but instead found a large amount of deer, rabbit, and small rodent remains. In the highlands of Mexico, Paleo-Indians hunted jackrabbits, gophers, and rats, as well as antelope. In eastern North America fish remains were found at the Shawnee-Minisink site, Pennsylvania. At other Clovis localities certain stone tool types suggest use in plant processing—even cultivation—as part of the Paleo-Indian cultural profile. In short, the stereotype of these early peoples as they appear in familiar illustrations of huge animals, rearing in fright and from the pain of spear wounds, surrounded by frenzied, skin-clad, hunters, is only part of the picture.

As for other aspects of Paleo-Indian life, we know almost nothing. Some half a dozen Clovis caches, discovered by chance

ABOVE: *Monte Verde, Chile, a flatland along a small creek now largely devoid of trees, was once a damp forest where Paleo-Indian hunters stayed for a cycle of seasons. Beneath a peat bog here, archaeologists unearthed extraordinarily well-preserved remains, including a hunk of mastodon meat.* OPPOSITE: *The dating of a fragment of mastodon tusk, once used as a tool, helped establish the average age of Monte Verde's occupation at 12,500 years ago.*

at various places in the West, contained what appear to be very special offerings or simply stashes of Clovis points and blades and other objects of bone and ivory. Most of these cached points are unusually large and of superb workmanship. One group was found associated with red ochre, suggesting that it may have been a religious offering—not surprising when one considers the deep roots of shamanism in the ancient hunting cultures of Siberia.

MANY DISCOVERIES OVER THE PAST CENTURY AND A HALF suggest that Clovis hunters were not the first Americans, but that others preceded them, perhaps by thousands of years. Such sites have been found from Canada to Chile, and their names form a litany of discouraging "almosts" in the archaeological literature—Lewisville, Texas; Tlapacoya and Valsequillo Reservoir, Mexico; Fort Rock Cave, Oregon; Wilson Butte Cave, Idaho; Pikimachay Cave, Peru; and on and on. For various reasons, most archaeologists view the early dates at these and other places with feelings that range from cautious skepticism to outright disbelief.

One of the best cases for pre-Clovis occupation of North America comes from James Adovasio's excavations of Meadowcroft rockshelter, Pennsylvania. There, a reliable sample of dates indicates human occupation between 12,275 and 8000 B.C. However, even these early dates no longer appear so unusual when one considers what Thomas Dillehay has found in the layers of Monte Verde, Chile.

Monte Verde, located some 20 miles inland, and about two-thirds of the way south along Chile's long stretch, lies by a creek in the shadow of a snowcapped volcano. Before the bulldozers cleared it for a housing development, the site lay in a cool, misty rain forest over an old peat bog, an environment that helped to preserve things virtually unknown at most early sites— wooden stakes, probably to secure hide tents; remains of medicinal and food plants; and even a child's footprint. Even more astonishing were the dates of these discoveries, which average around 10,500 B.C.—not only preceding the Clovis presence in the Americas, but also occurring in a location near the

The land along the Savannah River holds some of the most important archaeological sites in the southeastern United States, ranging from Paleo-Indian times to the period of European

contact. Among them, Stallings Island, near Augusta, Georgia, and Rabbit Mount, farther downstream,
have yielded pottery dated as early as 2500 B.C., the earliest known for all of North America.

end of South America farthest from the Bering Land Bridge.

The dates for Monte Verde have not met with unanimous approval. Archaeologist Stuart Fiedel, in particular, questions Dillehay's documentations of artifact provenance and, by extension, the early date proposed for the site. The controversy only underscores the fundamental problem of dating the earliest Americans with total certainty. There are other questions as well: How did the earliest arrivals come, by boat or by foot, or both?

All we do know is that beginning some 12,000 to 15,000 years ago the first of the gradual movements of people began eastward from Siberia across the land bridge into the Americas. We also know from new research in molecular and DNA sampling from both ancient and modern bone that this early wave of population was merely the beginning of at least three, perhaps six, separate major movements, and that among the latest arrivals were the ancestors of the present-day Inuit. We also know from the geologists that the timing of these movements was controlled by the great cadence of successive cold and thaw that in turn exposed and submerged the land bridge, while simultaneously blocking and unblocking the routes to the south. And we know that by the final disappearance of the Ice Age, most of the habitable parts of the Americas held growing numbers of people working to adapt to the changing conditions of their natural settings.

By about 8000 B.C. changes in climate were melting the northern ice to expose great areas of new land, and with these changes came others. The mammoths, mastodons, and other large animals that helped provide food for the Paleo-Indian hunters slowly vanished, to be supplanted in importance by smaller forms—deer, elk, and the modern bison, a descendant of the Ice Age buffalo.

FOR THE SAKE OF CONVENIENCE, archaeologists call the period of some 7,000 years, from the end of the Ice Age until the centuries around 1000 B.C., the Archaic Period. The hallmark of this long span is great cultural variety and versatility, as life in general became gradually more settled with the successes of adaptation to a variety of natural settings—along with increasing cultural complexity. As population continued to

BELOW: *Unearthed from the dry deposits of Lovelock Cave, Nevada, a canvasback drake decoy carefully made of brush, feathers, and paint suggests the innovations of the Archaic inhabitants of the region some 2,000 years ago.* OPPOSITE: *A series of Anasazi handprints marks a sheltered sandstone wall at Canyon de Chelly National Monument, Arizona.*

increase, the possibility of simply moving to unoccupied land for resources lessened, forcing people into more restricted territories.

Danger Cave, Utah, a wide-mouth grotto where people lived sporadically from about 8300 B.C. to the recent past, stands as one of the principal sources of our knowledge about Desert Archaic culture. Because of the extremely dry setting and the long occupancy, Jesse Jennings, who excavated the site, found many items of wood, bone, and basketry that survived there, along with sandals, bags, and cloth. In addition, Danger Cave yielded more than a thousand milling stones and hundreds of manos, or hand stones, used to grind seeds into meal—all of this in a series of occupation layers that eventually reached a depth of 13 feet.

Subsistence in the desert was not a simple matter. The remains of food in Danger Cave indicated a highly varied diet that ranged from mountain sheep to wood rats, and from pickleweed to charred grasshoppers. In short, the Danger Cave people typified the Archaic way of life—making the maximum use of whatever is available. Remains at Eva, in northern Tennessee, and countless other sites of the southeastern Archaic peoples along the rivers and coastlines from the Middle Atlantic to the mouth of the Mississippi, are marked by huge refuse heaps of discarded mussel shells. Many other

The roaring Virgin River cuts deep into sandstone at the gorge of the Narrows in Zion National Park, in southwestern Utah. The desert plateau in the region, including the Grand Canyon of the Colorado River to the south, witnessed the rise and decline of a succession of cultures dating from the end of the Ice Age.

settlements of the time, such as Koster, Illinois, thrived in the midst of relative bounty in a fertile river valley teeming with wildlife.

Archaeologists see the Archaic Period as ending at different times depending on the place, and they mark it with the momentous innovations and changes that grew from the accumulated knowledge of millennia of trial and error.

For many areas, the most important innovation of all was knowledge of agriculture. Richard S. MacNeish's excavations in Mexico's Tehuacán Valley revealed one center for an important milestone in the development of ancient American agriculture—the beginning of maize cultivation between 5000 and 3500 B.C., evidently the culmination of centuries of accident and experimentation by the gatherers of plants and wild grasses.

By around 2000 B.C. many areas of the Americas stood on the threshold of momentous change. The cultivation of squash, beans, and all-important maize had long been under way in Mesoamerica and soon reached the Peruvian highlands to join the potato and other crops. And, based on the evidence of ceramic griddles, manioc (which leaves no archaeological evidence of its presence) was cultivated by peoples of Amazonia and the coastal regions of northern South America. Initial experiments with plant cultivation began in the North American Southwest as early as 1500 B.C. In the Southeast the Archaic period ends sometime between 2000 and 1000 B.C. with the appearance of pottery.

In all these places populations continued to increase, creating a need for more complex social organization and a political mechanism for the management of vital resources and the new idea that the local environment could itself be manipulated—at least up to a point. Increasing economic surpluses provided obvious benefits, but also created competitive settings where the seeds of states and nations would soon sprout amid agricultural hamlets. Because of these and other forces, the accomplishments of Indian culture and civilization would now begin to accelerate in various ways throughout the Americas.

In the meantime, those who could not farm by reason of natural setting or cultural choice, among them the peoples of the polar regions, continued in the old and effective Archaic Period patterns of hunting, fishing, and gathering—and virtually transformed those activities into an art form.

CHAPTER 2

THE POLAR REGIONS

INDIAN AND ESKIMO PEOPLES
ADAPT TO THE DEMANDS OF LIFE FROM
BOREAL FOREST TO ARCTIC SEA

T HE WORD ESKIMO is but one of many unfortunate cases in which a wrong name stuck too long and became part of history. The word is an old Algonquian term of contempt meaning "eaters of raw meat," and it was allegedly voiced by Indians (another wrong name) of the subarctic region referring to their Arctic neighbors, whom they clearly disliked. In reality, the peoples we know as Eskimos call themselves Inuit, meaning something like "the people," although the term "Eskimo" is almost always used to describe the ancient way of life of early American Arctic peoples. However this history of name-calling might have evolved, the example involves two distinct peoples, both hunters and gatherers of great skill and inventiveness, whose related stories comprise the greater human epic of extreme northern North America.

The story of the Indians of the subarctic region began as early as 8000 B.C., when the great northern ice sheet had gradually shrunk to mere remnants of its size at the peak of the Ice Age, to be replaced first by glacial lakes and tundra, then, as the land continued to change, by a pioneer forest of spruce, pine, birch, fir, poplar, and, closer by the tundra, alder. This became the boreal forest that eventually covered most of inland Canada and Alaska. With the growth of the forest came caribou, moose, elk, bear— and Archaic Period hunters, probably descendants of early groups out of southwestern Alaska, as well as various Archaic groups who moved north into the area.

In the Aleutian Islands archaeologists have revealed early evidence of another aspect of ancient American pioneers of the north. At Anangula Island, one of the eastern Aleutians, flaked points, blades, and cores resembling those made in Siberia, along with stone bowls and scrapers, reflect a successful adaptation to coastal life, which clearly included the knowledge of seagoing boats by about 6000 B.C.

Archaeologists know very little of the long culture history of the subarctic peoples, undoubtedly Indian, who lived in the interior, except that their skills in hunting, fishing, and gathering were apparently as efficient as they were unspectacular. By about A.D. 500 the various subarctic groups had gathered into two broad language groups—the Athapaskan speakers to the west, and the Algonquian speakers of eastern Canada.

From very earliest times onward, however, these two great

The Kongakut River in northern Alaska's Arctic National Wildlife Refuge wilderness flows through a treeless mountain valley on its way to the Arctic Ocean. The terrain typifies a region where the cold year-round climate and relative

lack of vegetation limits or prevents the kind of natural soil buildup so common in temperate and tropical regions.
Thus, the rarity of stratified archaeological sites in the Arctic hinders efforts to date accurately most ancient remains.

subarctic Indian groups shared many characteristics. First, they were continually moving—at least in restricted areas—according to the availability of game and with the contraction and expansion of the forest boundary, for the northern climate continued to fluctuate even after the Ice Age had passed. Second, this movement, particularly in relation to the northern tree line, brought them into continual contact—and contention—with the other major group of peoples of the far north—the Arctic Eskimos.

THE ANCESTORS OF THE ESKIMO PEOPLES came relatively late to the American scene, and much of what we know about their arrival comes from a remarkable site. Onion Portage, named for a profusion of wild onions that grow in the vicinity, lies within a meander loop of the Kobuk River, about a hundred miles inland from Alaska's west coast. Discovered in 1941 by J. Louis Giddings, and later excavated by Douglas Anderson, Onion Portage is that rarest of the rare in Arctic archaeology—a site with stratigraphy. At Onion Portage Anderson found lots of it—some 20 vertical feet in all, holding some 50 different levels of remnants of human occupation that took place intermittently over a period of thousands of years, beginning in Paleo-Indian times and ending only a few centuries ago. Among the layers of this remarkable site, some higher up in the sequence revealed much of what we know of the earliest Eskimos.

Beginning about 2200 B.C. these first Paleo-Eskimos appeared in northern Alaska from the west. At Onion Portage and other sites in the region, their remains appear in the form of small delicately flaked stone points, knives, scrapers, and other tools. Based on the nature of these implements and their widespread distribution, Arctic archaeologists see this class of implements as the hallmark of what they have dubbed the Arctic Small Tool Tradition. By using the occurrence of this distinctive set of tools and its

LEFT: *Seal hunters used this lure, carved of wood and inlaid with seal claws, to scrape the surface of the ice near a blowhole. The resulting rasping would often draw the unwary seals into harpoon range. The artifact came to light near Point Barrow, Alaska, at the site of Utqiagvik, in the frozen ruins of an Inupiat house destroyed by a storm some 500 years ago.*
BELOW: *A traditional Inupiat berry basket was crafted from wood.*
OPPOSITE: *An Eskimo mask of caribou skin and fur, together with the other artifacts, reflects the variety of Eskimo subsistence patterns ranging from the quest for sea mammals to the land-based patterns of gathering and the pursuit of caribou.*

appearance at various sites as clues to ancient Eskimo settlement and movement, archaeologists have revealed one of the most remarkable stories of human migration and movement in the archaeological record of the Western Hemisphere—remarkable not only for the great distances involved, but also for the nature of the land where it took place.

Measured roughly along the Arctic Circle, the tundra stretches for some 3,000 miles between the Bering Sea and Greenland. For those accustomed to more temperate places, which include virtually everywhere else on Earth, it seems a cruel and harsh place for continued human existence—and it is. The far north is a place of seemingly eternal winter where the oceans remain frozen for

The Noatak River, in the Brooks Range of northern Alaska, makes its way to the Chukchi Sea via Sevisok Slough. A meandering river such as this snakes across its flat alluvial valley like a sidewinder, a movement that creates ever larger

curves, leaving behind concentric bands on the land as traces of the river's progress. In the Arctic as elsewhere, such wandering waterways thus obliterated the earliest traces of any human culture that may have been present.

eight months a year. It is a bitterly cold and treeless place where temperatures can plummet below minus 75° F.

In central Alaska the tundra receives little precipitation, and what does fall cannot penetrate the frozen ground, so the landscape is one of standing water and tufts of cotton grass, lichens, mosses, and bizarre miniature "forests" of birch and willow. Along the coast of the polar sea, to the north beyond the Brooks Range, stretches the true Arctic tundra, a vast treeless plain carpeted by reindeer moss that draws caribou in their annual migrations from one feeding place to another.

The seas that border this land provide the Arctic coastal lands with marine life of unusual variety and richness—walrus, sea otter, seals, and whales—and have given a unique and special character to the lives of the hunters of the high Arctic. The Paleo-Eskimos, separated from their Aleutian relatives by about 2000 B.C., began the first of three great movements to the east. The first involved the early Eskimos, known to archaeologists by the presence of the Arctic Small Tool Tradition, who eventually completed a transcontinental journey over tundra, water, and pack ice—a journey that ended on the northwest coast of Greenland.

During the centuries of their spread across the entire North American Arctic, the ways of these migrants gradually changed, particularly with regard to food procurement. In the beginning the early Paleo-Eskimos hunted the polar bear and other land animals for protein and fat so vital in the Arctic; by 1600 B.C. the hunters were taking increasing advantage of the seal, walrus, and other sea mammals. Archaeological remains of this period generally lack stone whale-oil lamps, indicating that these people depended mainly on the scanty supply of wood as fuel for both light and heat.

Even as this movement was taking place, each region along its path witnessed its own local developments. Between 1600 B.C. and the beginning of the Christian era, numerous local versions of Eskimo culture emerged at various times and various places above the tree line from Alaska to Greenland.

One of these expressions of the early life of Eskimo hunters may be seen in what archaeologists found at Ipiutak, an ancient settlement near Point Hope on the northwest coast of Alaska, a thriving community of Arctic Ocean hunters during the first five centuries A.D. Ipiutak, the largest known prehistoric Eskimo site,

OPPOSITE: *Bearded seal on an Arctic ice floe represents a species that lives throughout the world's Arctic zone. Along with whales and walrus, land animals such as the musk ox and caribou form the main objects of seasonal hunting by both ancient Eskimo and modern Inuit peoples.*
ABOVE: *The bow and arrow came into use as early as 2000 B.C. among pre-Dorset groups. These hunters use the recurved bow characteristic of the Central Eskimos. Slit goggles protected hunters against snow blindness.*

contained some 600 rectangular, semi-subterranean houses of wood arranged in four closely packed rows on beach ridges beside the Arctic Ocean. Ipiutak houses yielded numerous objects of everyday use, among them knife handles and harpoon sockets, many elaborately carved in delicate low relief punctuated with fine line work and dots—a style related to the Old Bering Sea art produced by the walrus hunters who dwelt along the eastern Siberian coast and nearby islands of the Bering Sea.

The Ipiutak cemetery held at least 138 burials contemporary with the Ipiutak houses, and, like the houses, the burials lay along parallel gravel ridges. Some of these internments were in rectangular log coffins buried in deep pits dug by shovels of whalebone, several of which had been left on the coffins. Others had simply been placed on the ground and covered with wood and sod. These surface burials apparently belonged to individuals of special status, including shamans—intermediaries between the natural and supernatural worlds. These held all manner of elaborately carved ivory—swivels and chains, daggers, arrowheads, openwork carvings of human heads and animals, and funerary masks.

Shamans—and the very word comes from one of the old native Siberian languages—were an important key to circumpolar

A panoramic view shows the sweep of the Thelon Valley, Northwest Territories, Canada, with Whitefish Lake in the background. Here, darker trees define the distinctive dunes and ridges that mark the landscape.

Those deposits of sand and gravel, known to geologists as eskers, were dropped by rivers and streams that flowed beneath the ice sheet that once covered most of the region during the last ice age.

life, fulfilling the tasks of religion and curing. The canons of ancient Arctic belief were probably in keeping with the present view among some peoples of the same region, who see the world as a place where people and animals are equal partners in a reciprocal arrangement—with one providing food in return for deep respect—and with formal rituals to emphasize that respect.

One of the more elaborate shaman burials was reported at the Siberian site of Ekven by Russian archaeologists S. A. Arutiunov and D. A. Sergeev. The burial, belonging to the Old Bering Sea culture, was enclosed by bones of a beluga whale and proved to be that of a woman. She lay on a wooden platform lined with stones, surrounded by implements and other objects of ivory, wood, shell, bone, and stone—including those customarily used by men and women. The presence of a drum handle and a wooden mask served well to reflect her great power.

IN THE MILLENNIUM BETWEEN 500 B.C. AND A.D. 500 the various cultures that extended across the Arctic had seen another great cultural development. Known as the Dorset culture, it emerged from among the early Eskimo peoples of eastern Canada and Greenland.

At the peak of their expansion in the centuries before the beginning of the Christian era, the Dorset Eskimos occupied places as far south as southern Newfoundland and Labrador, but by A.D. 1000 they had retreated to northern Labrador, the central Canadian Arctic, and northwest Greenland.

The Dorset Eskimos were mainly hunters of marine mammals, especially seals and walrus, as their tool inventory shows.

LEFT: *Carvings of baby seals, perhaps ornaments, reflect the rich ivory-carving tradition of the Eskimos. The raw material, a by-product of the hunt, served for objects of both art and utility, and all reflect the distinctive styles of their dates and places. For example, the hunters of the Old Bering Sea culture of the first millennium B.C. decorated the various ivory parts of their complex harpoons with extraordinarily delicate patterns of engraved lines and dots. The same style marks ivory needle cases, sinew twisters, knife handles, as well as other household items.*
OPPOSITE: *This intricately carved composite funerary mask, perhaps that of a shaman, was found disassembled in a 1500-year-old burial at the Ipiutak site, Point Hope, Alaska.*

Strangely, they did not use the bow and arrow but only harpoons and spears, along with many of the tools and gadgets that survive in use to the present time—including snow knives for making igloos, ivory sled runners, and soapstone lamps. Although they must have used skin boats, information about them has remained elusive in the Arctic record.

By around A.D. 900 another warm period came to the Arctic, fostering great changes in culture and in the movement of peoples. The principal group involved in this drama we know as the Thule Eskimos, who evolved in Bering Strait and northern Alaska, then expanded east, into the Dorset area. Equipped with everything from kayaks to the larger umiaks, and with advanced harpoons, dogsleds, and the bow and arrow, Thule culture stands as the culmination of Arctic technology. In their advance from Alaska to Greenland, Thule peoples completely overran, and sometimes blended with,

Some 500 years ago the Inuit settlement of Qilakitsoq, "the sky is low," lay on the narrow shore of an inlet on Greenland's west coast, 280 miles north of the Arctic Circle. In this reconstruction of a summer scene,

men butcher a narwhale while a woman gathers bird eggs from a cliffside nest. In winter, the people will abandon their skin tents and move into the sod-covered dwellings at the base of the cliff.

45

OPPOSITE: *Bear tracks lead to water in north-western Alaska's Noatak National Preserve wilderness. They are an unmistakable mark of the grizzly, who shares his western Arctic habitat with the polar bear, whose range spans the northern Arctic shores from Siberia to Greenland. Both species provided food for Arctic hunters, whose skills on land and sea were honed to perfection through the 7,000 years or so that they occupied the frozen reaches of the far north.*
RIGHT: *A beluga whale, quarry of the sea hunt for both the Eskimos and the Indians of the northern Pacific coast, surfaces in the icy northern waters.*

the Dorset peoples whom they encountered. With their inventory of boats and specialized tools and weapons, the Thule utilized virtually every resource the Arctic had to offer, on both land and sea. By A.D. 900 they hunted whales and seals from Baffin Island to southern Labrador and from Hudson Bay to Greenland.

By this time, too, the same warming trend that made the hunt for sea mammals and caribou so successful for the Thule had also drawn Norse settlers from the east to Greenland—the first Europeans to be seen by any Native Americans. The Norse established cattle farms in the lands around southern Greenland's coastal fjords. By about A.D. 1350, when the Norse farms were proving less and less productive, Thule Eskimos had appeared at the mouths of waterways, where they hunted sea mammals. In the cold period that soon ensued—the time known as the "little ice age"—the European farms finally failed, and the Norse abandoned their Greenland settlements. The lesson of their demise seems clear. In contrast to the peoples of the American Arctic and the more than a hundred generations of experience and success that spawned them, the rigid patterns of the European agricultural system of the time could not, and would not, adapt to the environment of the region.

In the aftermath of the Norse departure from America the Thule hunters inherited the position of sole proprietors of the North American Arctic until the arrival of European whalers and explorers in the 19th century.

CHAPTER 3

THE WEST

FROM THE DESERT WESTWARD TO THE SEA, HUNTERS, FISHERS, AND GATHERERS SUCCEED IN A WORLD WITHOUT FARMING

ARLY IN THE MORNING OF AUGUST 29, 1911, near Oroville in Northern California, a group of slaughterhouse butchers awoke suddenly to the barking of dogs. Investigating, they found a man, emaciated and naked save for a tattered poncho of wagon canvas, crouched at bay in the corner of a corral. As his biographer Theodora Kroeber later put it, "he was at the limit of exhaustion and fear."

The Oroville sheriff was summoned and, not knowing what else to do, held the man in the county jail, in a cell for the insane.

At the time of the episode, Kroeber's husband, Alfred, an anthropologist at the University of California, Berkeley, saw the sensational press notices that appeared in the wake of the episode and realized, as had the Oroville sheriff, that the starving "wild man" was almost certainly a northern California Indian. Other California Indians who had tried to communicate with him could not understand the language he spoke. However, one of Kroeber's linguist colleagues, Thomas Waterman, armed with vocabularies from the university's archives on local Indian languages, haltingly interviewed the man as best he could, and slowly the story of the "last wild Indian in the United States" began to emerge.

He was the lone survivor of a small band of Yahi Indians, a subgroup of the Yana, who had lived in a part of the northern California wilderness thought by the white settlers to be far too remote and desolate for settlement and development. Members of his small band, including his mother, had apparently died following an encounter with a party of white surveyors three years earlier. His short, burned hair was his sign of mourning.

The man, as was customary among the Yahi and others, never revealed his name to the strangers who took him in, but instead accepted the name Ishi, which in Yana means "man." The Bureau of Indian Affairs, which had legal jurisdiction over this unique case, granted approval for Ishi to live out his life in the private guest wing of Berkeley's new Museum of Anthropology in San Francisco. Kroeber and Waterman—both of them now close friends with Ishi—along with students and others in and around the campus, helped him live free of most of those who sought to intrude into his life.

Ishi willingly shared much of his knowledge with his hosts— his ways of making bows, stone arrow points, wooden salmon

PRECEDING PAGES: *Spruce, rose, and paintbush mantle the rocky windswept coast of Vancouver Island, British Columbia, home of the Nootka and other Northwest coast peoples. Helped by relatively warm currents that cross the North Pacific from Japan, the near-shore waters of these high latitudes held an extraordinary variety of life, as did the inland rivers. Even the earliest cultures of the region lived mainly, and successfully, by means of riverine and maritime resources.*

OPPOSITE: *Against the backdrop of the snow-clad Sierra Nevada, a complex of enigmatic petroglyphs covers a rock surface at the western edge of the desert country of the Great Basin, near the California-Nevada border. The setting here, home of the modern Paiute and Shoshone peoples, provided a less secure living than the coastal lands, but one that sufficed for hunters and gatherers.*

harpoons, and other tools; his use of deer-head decoys, of otter skins for making quivers; the nuances of his language, the Yahi dialect of Yana; his knowledge of plants for food and curing; and his stories—tales of the powerful hero-hunter Wood Duck Man and his amorous adventures; accounts of the rhythm of the days and the moons that marked the round of the seasons; and of his everyday life and labor as a hunter and gatherer in a place then untouched by foreigners.

Ishi died in 1916, a victim of tuberculosis—one of the many diseases brought to his land by the whites.

Like the Yana and other California groups, many of the ancient peoples west of the Great Plains and Rocky Mountains never farmed—some because arid climate made it impossible, others because the favorable conditions of terrain and climate created a natural bounty that made agriculture irrelevant. Thus, while those living in the Great Basin region generally continued the lifeways of their Desert Archaic ancestors, the ancient Californians thrived as a veritable mosaic of distinctive cultures in a patchwork of varying natural settings, each of extraordinary productivity.

THE CHARACTER OF CULTURAL DEVELOPMENT throughout the time span between about 2000 B.C.—the end of the Archaic period—and the arrival of the first Spanish colonists in 1769 comes from a multitude of archaeological sites ranging from the interior mountains and valleys to the coast, and eastward into desert and prairie country, and from tiny campsites to elaborate settlements. These formed a mosaic of differing cultures and languages, each of which occupied a small and distinct microenvironment. Throughout the period the fundamental pattern of culture all over California was one of adaptation to whatever the local setting had to offer. Such patterns do not necessarily produce lasting works of great art and architecture, but they do reflect great ingenuity in living life in a manner that guarantees a continuing and varied food supply from a sustainable setting.

The ancient Californians didn't use clay pottery, but baskets instead. Not just ordinary baskets, these skillfully woven containers held not only seeds and other foodstuffs, but water as well. Cooking was accomplished by dropping heated clay balls into the water until it boiled. Many of these baskets doubtless were objects of fine art as well. Anyone who sees a modern California Indian basket, say a Pomo example, so incredibly tightly woven and adorned with tiny shells and bright feathers, can hardly fail to feel a great loss at those examples that have vanished along with the anonymous artists who made them.

The sheer variety of plants and other resources available to ancient peoples of California fostered strong patterns of trade, founded on cultural traditions of storage of local goods and their regulated distribution through trade arrangements among the various peoples. Thus did California support the highest density nonfarming population in all of North America in the days before the havoc of land destruction and disease that began after contact with whites. Of the estimated 300,000 inhabitants in California in 1769, only a relative few remained a century and a half later.

The most vivid description of life as it might have been in ancient California is that of writer Malcolm Margolin, who imagined the land during a spring and summer several hundred years ago:

"On the great wind-riffled prairies Pomo and Wintun women stoop under the burden of pack baskets to dig roots from

Lake Tenaya, at Tuolumne Pass in Yosemite National Park, a main passage through the Sierra Nevada, connects two disparate environments—the desert country of the Great Basin and the more fertile area of California to the west. The

lake name memorializes Tenaya, or Tenieya, leader of a group of some 350 Miwok, Yokut, and Chowchilla Indians who surrendered to the California militia after being hunted down for attacking white miners who had invaded their lands.

April's emerald landscape, or beat ripe seeds from the tawny grasses of August. In the oak-studded foothills, listen to the monotonous thump of stone pestles wielded by patient Maidu women grinding acorns. Down in the delta marshes and sloughs of San Francisco Bay a Yokuts woman uproots thick aquatic tubers while guiding her tule raft silently along many miles of curving, green waterways.

"High in the conifer-spired Sierra, an agile Miwok climbs a tall and stately sugar pine to snap off the pendulous cones, while family members wait eagerly below to gather the pine nut delicacy. In the corrugated basin-and-range landscape of eastern California a small Washoe family cluster, weighted down with burden baskets, move on their annual trek up the Sierra's steep eastern escarpment to summer fishing and gathering encampments near thawing mountain lakes."

WHILE GENERATION UPON GENERATION of California Indians pursued their varied lives, others, living along the Pacific between California and the Alaska Panhandle, followed their own collective destiny. The late 18th-century European explorers encountered them as the hunters of sea mammals, the historic ancestors of the present Haida, Nootka, Kwakiutl, Tlingit, and other groups. These Northwest Coast cultures are famed for both their woodcarving skills and their art style—a stunning combination manifest in ritual objects ranging from the famed totem poles of the Queen Charlotte Islands to the formidable seafaring canoes used in the pursuit of whales. Northwest Coast art

OPPOSITE: The Antelope Valley of southern California, its slopes mantled in poppies, goldfields, and owl's clover, hints at the beauty and diversity of the region's flora. California plants include more than 5,000 kinds of native ferns, conifers, and flowering plants, and a variety of forms that range from fragile herbs to giant sequoias—a diversity unmatched in most other areas of equal size. Indians of the area utilized this extraordinary bounty as sources of medicines, raw materials for the making of textiles and baskets, and many other purposes.
ABOVE AND RIGHT: *In their fineness and intricacy of design, Pomo baskets reflect the enormous talent of their makers. The traditional Indian baskets of California, though differing from group to group, served from ancient times onward as containers and cookware and as masterpieces of aesthetic craftsmanship.*

LEFT: *Two bald eagles in aerial contest represent one of two varieties of the bird native to California.* OPPOSITE: *The rugged coast of Big Sur marks the edge of the former homeland of the Costanoan Indians. Numbering an estimated 7,000 at their peak in prehistoric times, this group occupied most of the coastal lands south of present-day San Francisco Bay—much of it made up of rugged mountains and steep canyons. Costanoans—their name derives from the Spanish word for "coastal dwellers"—subsisted mainly on acorns, salmon, and oysters. White settlement in their lands during the late 18th century disrupted their lives and their culture. By the time of the census of 1910 they had become extinct.*

is also known through smaller objects of ritual and household use—ceremonial boxes, masks, rattles, drums, and whaling hats.

Archaeologists trace the ancestry of the Northwest Coast peoples back to the Archaic hunting cultures of the region. Much of the scholarly knowledge comes from the occupation layers of the Cape Flattery site, on the United States side of the Strait of Juan de Fuca. According to the evidence at Cape Flattery, at around 2000 B.C. hunters in the interior pursued deer, elk, and smaller mammals. For fishing in both the rivers and ocean they utilized skin boats and large dugout canoes. All these adaptations to the cool, moist forests and the adjacent sea, along with increasing skills in woodworking, gradually led them to form large villages of plank houses, some of which covered several acres. By A.D. 400 Northwest Coast cultures were evolving rapidly into the Tlingit and others.

All these groups soon became oriented primarily to the sea, drawn by the bounty provided by the warm North Pacific Current that had so helped their early neighbors in the Aleutian Islands and on the nearby Alaska coast—cod, herring, halibut, and salmon, along with whales, sea otters, and seals. Among the Northwest Coast peoples to take advantage of this setting were the Makah. They, like their neighbors the Nootka, were great whalers, and their beachfront village, Ozette, strategically located near a whale migration route and some 15 miles south of Cape Flattery, was for many generations one of the main whaling centers south of Alaska. Sometime around A.D. 1500 disaster struck Ozette when heavy rains triggered a mud slide on the steep slope behind the settlement. Though the people apparently escaped, the village was partially buried—and the

Makah hunters of the coastal waters off present-day Washington State's Olympic Peninsula
return to their home settlement of Ozette in seagoing dugout canoes with a

gray whale in tow. This reconstruction is based on remains of Ozette's houses and other artifacts miraculously preserved by a mud slide some 500 years ago.

Makah soon built a new Ozette nearby. In 1970 winds and waters of a severe winter storm pummeled the coast of Washington State, and before it ended, artifacts and parts of a plank house from the old village lay exposed for the first time in nearly 500 years. The unique discovery soon triggered one of the largest archaeological efforts in Northwest coast history. In it, Richard D. Daugherty of Washington State University joined forces with the Makah Cultural and Research Center and began a program of excavation and conservation that lasted for 11 years.

Eight to twelve feet below the surface Daugherty and his team discovered five houses, with their contents sealed and well preserved—much as the ash of Vesuvius preserved the finds at ancient Pompeii. Each house structure at Ozette measured 60 to 70 feet long by 35 feet wide, and each proved to be a treasure trove of information about life in the ancient Northwest.

As the late writer Gene Stuart, who was there to witness some of the work, described it, "By 1979 some 80,000 recovered objects, from baskets and whale bones to halibut hooks and looms, spoke of everyday life more than 450 years ago. Wood chips marked a workshop area. A sleeping mat still lay spread on a bench.

"A ranking family had lived in the rear quarter of one house—their ceremonial gear, whaling harpoons, and a distinctive cedar-bark hat worn only by the elite all indicate high status. A large whale fin carved from cedar and inlaid with 700 sea otter teeth may have been important in ceremonies. The design may depict a mythical thunderbird that hunted whales and carried them in its talons to its mysterious mountain home."

Perhaps the ultimate symbol of the Northwest Coast peoples

OPPOSITE, UPPER: *An ornate halibut hook of elaborately carved wood with a bone barb was crafted by the Tlingit of the southeast coast of Alaska.*
OPPOSITE, LOWER: *Another special object is a Haida ladle made of horn and shell.*
RIGHT: *A remarkable discovery from Ozette— a cedar carving of a whale's dorsal fin, studded with sea otter teeth—was probably an object for display at rituals and ceremonies connected to the sea hunt of the ancient Makah.*

is the totem pole, a creation unique on Earth. Each of the imposing carvings, at once a work of high art and complicated meaning, served as a monumental symbol of status and ancestry for a family wealthy enough to commission it. The symbolism worked into totem poles includes heads of animals and humans, both mythical and real. Some of these represent one or more crests inherited by a specific family. Others may be "read" as family histories, telling how the rights to certain crests came to be owned through encounters with supernatural beings. Still others refer to successful competitions over rival chiefs.

Totem poles also stand as appropriate symbols of the world of the Northwest Coast peoples—a world inhabited by human, animal, and supernatural spirits who can transfer from one state to another; a world of elaborately masked shamans and dancers performing the dramas of myth and history; and a world where status could be maintained by the potlatch, or gift ritual, in which

Skidegate, a Haida village in the Queen Charlotte Islands, held a remarkable display of totem poles fronting the sea. Villages such as this dotted the Northwest coast until the early years of the 20th century. The photograph, taken by

John R. Swanton, later a renowned ethnologist of the Smithsonian Institution, exemplifies the context and number of great Northwest Coast carvings that stood until their removal to museums and collections throughout the world.

Carved poles such as this Tlingit totem pole at Sitka National Historical Park, Alaska, have come to stand as the main hallmark of the art of both the ancient and present-day peoples of the Northwest coast. Such carvings stood outside dwellings, objects of respect and awe intended to convey timeless statements of family history, myth, and the social status of their owners and their carvers. Northwest Coast carvers perpetuate the tradition today, helping to maintain the cultural identity of their people.
RIGHT: *This superbly carved wooden rattle, dominated by the image of all-important Raven, once reflected the high status of its owner, a Tlingit chief.*

rare and costly objects such as ornate copper plates were destroyed, given away, or willingly consigned to oblivion in the waters.

Many of the old totem poles that once graced the towns of the Haida, the Tsimshian, and others of the Northwest coast have been moved to museums. In many cases, the only reminder of their original context lies in the rare and occasional photographic images made in the late 1800s, when the poles still stood proudly at places like Skidegate, Kitwanga, and Masset, British Columbia. The culture they represent, however, is still alive and well in the hands of talented descendants of the Northwest Coast tradition such as Robert Davidson, a Haida, himself the great-grandson of the artist Charles Edenshaw. Davidson, through his enormous talent in sculpture, woodcarving, painting, and print-making—along with a profound understanding of his heritage—is a main force in the rebirth of the traditional art of his people and of the Northwest coast in general. In 1969 he carved the first totem pole in living memory to be erected in his hometown of Masset.

As for Ishi, his story did not end with his death in 1916, for while his body had been cremated in accordance with his wishes, his brain had not. Instead, it had been saved, preserved for study, and sent to the Smithsonian Institution in Washington. On August 8, 2000, in a ceremony there, Smithsonian officials formally returned the brain to a group of California Indians descended from the Yana. Three weeks later the organ was reunited with Ishi's ashes, and all were buried in a secret ceremony near the base of Mount Lassen, ancestral home of the Yahi. By virtue of this rite Ishi completed the grand cycle of the body and spirit and thus stands as perhaps the greatest survivor of all.

CHAPTER 4

THE SOUTHWEST

PEOPLES OF A MAJESTIC BUT ARID LAND ACHIEVE GREATNESS AMID DESERT PLATEAUS, MESAS, AND VALLEY OASES

W HILE SOME OF US THINK OF THE AMERICAN SOUTHWEST as a barren, hilly desert studded with giant cactus such as forms the backdrop for old Western films, that stereotype applies to only a relatively small area in southwestern Arizona and adjacent Mexico. The rest of the region holds a great variety of low basins, rugged mesas, deep canyons, shallow washes, mountains, desert plateaus, and plains. Rainfall varies from almost none in the desert basins to the low side of moderate in the higher mountain and mesa country—and it is erratic in all parts of the Southwest.

Evidence from Ventana Cave, Arizona, and other Archaic period sites shows that by about 2000 B.C. Southwestern hunters, gatherers, and foragers had successfully adapted to their arid environment—this by creating a way of life much like that documented for their contemporaries at Danger Cave, Utah, and elsewhere in the Great Basin. The introduction of agriculture, an invention apparently diffused from earlier Mesoamerican centers to the south, took place slowly and erratically beginning about 1500 B.C. Armed with the knowledge of the cultivation of maize, beans, and squash, and, after about A.D. 200, the use of pottery, the various peoples of the Southwest gradually evolved cultures of great distinction and accomplishment in the arid land.

Three main peoples—the Hohokam, the Mogollon, and the Anasazi—took part in the story of the ancient Southwest between the early centuries of the Christian era and the coming of Spanish explorers. Their collective accomplishments occurred at different times and in different parts of a large area that includes southern Utah and Colorado, most of Arizona and New Mexico, and large portions of the Mexican states of Sonora and Chihuahua.

THE HOHOKAM LIVED IN THE DESERT of southwestern Arizona, mainly in the valleys of the Gila and Salt Rivers. In the language of their modern O'Odham descendants, the name means something like "that which has perished." Another O'Odham word, *Shoaquick*, names the principal Hohokam site—Place of the Snakes—or simply Snaketown.

Snaketown takes up some 250 acres of gently rolling desert beside the north bank of the Gila River, some 30 miles upstream from its juncture with the Salt. Today the Gila is little more than

PRECEDING PAGES: *The jagged form of Ship Rock rises majestically from the desert of the Colorado Plateau in northwestern New Mexico. According to Navajo legend, the 2,188-foot formation is the remains of the great bird that brought the Dineh, or the People, as the Navajo call themselves, to the North American Southwest from their ancestral home in the western subarctic region of present-day Canada. Evidence from their ancient sites in the Southwest indicate the Dineh arrived in the region around A.D. 1500, joining the Anasazi, already in the area.*

OPPOSITE: *A painted bowl in the Classic Mimbres style of the 11th-century Mogollon people who lived in present-day southwestern New Mexico depicts an anthropomorphic bird, perhaps a masked fisherman or a supernatural being, with a successful catch.*

Shaded from a parching sun by cottonwood and willow, the San Pedro waters a narrow oasis in Arizona's Sonoran Desert. The river lies along the eastern lands of the ancient Hohokam people.

The Hohokam, noted for their skilled work in building great systems of canals and locks,
were the first farmers to irrigate successfully the desert country of southern Arizona.

a dry, windblown riverbed in the midst of arid desolation. Before modern development took its toll, however, the river flowed constantly, and in ancient times its water was the key to North America's first successful conquest of the desert by irrigation.

We will probably never know the full extent of the Hohokam system of irrigation canals now buried beneath the desert. Estimates run to 300 miles, but their extent matters far less than the fact that they were built at all, and worked. The Hohokam, by much trial and error, solved problems of silting and gradient by building catch basins and diversion dams.

Of ancient Snaketown itself, excavations by Emil W. Haury in the 1930s and 1960s revealed the hard-packed floors and post-holes of more than 200 houses. These were rectangular, oval, or circular in plan, depending on the fashion and custom of the time. The dwellings generally held a central hearth and walls and roofs of reeds—a combination that survived into modern times among the O'Odham.

Throughout much of the first millennium A.D., Snaketown watched its irrigated valley grow green every summer, and it thrived. The sculptors of Snaketown made striking stone bowls and flat, rectangular paint palettes, many in the forms of men or desert creatures; and the potters molded tiny slit-eyed figurines, along with painted vessels in the shapes of humans and animals.

The slit-eyed figurines are reminiscent of those made by the early farming peoples of Mesoamerica—so much so that Emil Haury was of the opinion, after a lifetime of studying the Hohokam, that the origins of these early masters of the desert may lie somewhere in the arid country of northern Mexico. Another feature discovered at Snaketown reinforces the possibility of a Mesoamerican presence—an earthen ball court in the form of a giant basin with a playing field more than 70 yards long,

LEFT: *A 1,000-year-old red-on-buff effigy vessel reflects the skill of ancient Hohokam artisans in the ceramic arts, particularly in their distinctive use of interlocking "positive" and "negative" scrolls and other geometric designs.*
BELOW: *Etched shells display an ancient decorative technique unique to the Hohokam. Their makers first applied natural pitch in the desired pattern, then immersed the shells in an acid solution, probably made from fermented cactus juice. As the acid ate away the exposed surface of the shell, the design appeared in raised relief.*

ABOVE: *Ornate carved stone paint palettes, some in the forms of animals, once held ochers and other mineral pigments, possibly for body painting.*

which vaguely resembles counterparts made of stone that existed at many of the great cities to the south. The builders of Snaketown and other Hohokam settlements also raised small platform mounds, perhaps another hint of a Mesoamerican connection.

Another Hohokam achievement—etching—intrigued archaeologists for years. It appeared on finely crafted shell jewelry bearing animal figures in low relief—evidence of what Haury called "a touch of the flamboyant." Such luxury objects had led scientists to the conclusion that the Hohokam craftspeople discovered the process of controlled etching with acid centuries before the Renaissance Europeans did. Final proof came with the discovery of an unfinished ornament—a white shell with a coating of acid-resistant pitch worked into the silhouette of a four-legged creature. The ancient artisan had applied the pitch but had never soaked the shell in the weak acid solution—probably derived from the fermented fruit of the saguaro—necessary to eat away the exposed surface and leave the protected figure in relief.

Many of the hallmarks of Hohokam culture began to disappear after the 12th century: The people of Snaketown scattered up and down the valley, and another kind of architecture—and perhaps another people—began to appear along the Gila.

"As large as a castle and equal to the largest church in these lands," reads the journal of Father Eusebio Francisco Kino, Spanish missionary to the Southwest and the first European to

Detail of a classic black-on-white Mimbres bowl shows a stylized bighorn sheep surrounded by the intricate geometrical patterns that characterize the style, which reached a peak between A.D. 1000 and 1150 in what is now southwestern New Mexico. Such painting, notes archaeologist Douglas W. Schwartz, "represents a powerfully inventive and expressive climax in the traditional Indian arts of the American Southwest. Using elegant lines and dynamic masses, the artists placed complex, non figurative, representational, and narrative compositions on the interiors of hemispherical bowls. The resulting paintings communicate authority, skill, and a remarkable perception of reality." Of the estimated 10,000 or so Mimbres vessels known, almost all bear "kill" holes in the bottom made before burial, as does this example, perhaps done to release the spirit of the piece.

observe the ruin of Casa Grande, some 30 miles upstream from Snaketown, near present-day Coolidge, Arizona.

Unique for its size and form, Casa Grande consists of a huge tan cube of layered, sun-hardened caliche within a compound surrounded by a high wall and some small outbuildings. Its ancient surface—smooth but marked by age cracks and featureless except for dark entryways and ventilation holes—hides a four-story complex of more than a dozen rooms.

The original function of Casa Grande remains elusive, as does the precise identity of its builders. Some see the monumental structure as an apartment complex, others as a fortress for defense, still others as both—a kind of 14th-century desert palace that marked the horizon with its prominence, proclaiming to all who gazed upon it the status and the invincibility of its occupants. Its builders are not known with certainty. Some see Casa Grande as a Hohokam work, while others attribute it to outsiders who entered the area around A.D. 1200.

THE NAME MOGOLLON refers to the peoples who lived in the mountains and low-lying lands east of the area inhabited by the Hohokam. That portion of the Southwest, now bisected by the United States-Mexico border, includes extreme southeastern Arizona, most of southern New Mexico, and parts of Mexico's states of Sonora and Chihuahua.

Mogollon artifacts and house sites dating before A.D. 700 look very much like those in the Hohokam area, but Mogollon remains after that date are clearly recognizable by the presence of their distinctive pit houses. These efficient semi-subterranean dwellings, relatively deep, with sloping entrance ramps, served as the residences of choice for the next 300 years or so.

From the pit house settlements the Mogollon made maximum use of nature's frugal yield. From the dry layers of caves in the Pine Lawn Valley comes evidence of use of the forest for implements and building as well as a source of nuts, seeds, roots, and berries. From yucca and reeds came the fiber for sandals and material for basketry and mats. Local clays provided raw material for the polished red or red-on-brown pottery that archaeologists recognize as one of the hallmarks of the culture.

After about 900 the Mogollon, perhaps influenced by

pottery styles among the Anasazi to the north, began applying a white slip to their pottery. A century later Mogollon potters in the Mimbres Valley of New Mexico began decorating the white surfaces of their bowls and plates with motifs rendered in black paint—some geometrical, many with life forms. Examples of Mimbres black-on-white ceramics from Mogollon burials show both the incredible range and utter beauty of this pottery—and the reason that many who gaze at the old vessels consider the paintings on classic Mimbres pottery as representing one of the greatest art styles ever developed in the ancient Americas.

Perhaps the most remarkable example of classic Mimbres bowls and plates came from the Swarts ruin in southwestern New Mexico, a cluster of 172 sunken, rectangular rooms excavated from 1924 to 1927. Of more than 700 intact or reconstructed black-on-white plates and bowls that accompanied burials beneath the dwelling floors, only a handful of the complex, delicately rendered geometrical motifs were repeated. Vessels decorated with life-forms depicted plants, animals, and insects. Among these are flowers, birds, frogs, lizards, fish, and rabbits—the last perhaps associated with the moon, for the Mimbreños—like many other native North American peoples—saw a rabbit, not a "man," in the moon. Among the insects and other small creatures depicted on plates are a grasshopper, a scorpion, and even an inchworm! Still others show humans, both alone and interacting with other people, animals, or birds.

Mogollon culture in the Mimbres and Pine Lawn Valleys continued until around A.D. 1300, by which time it had been diluted beyond recognition by centuries of the ever increasing influences of the expanding Anasazi, who at around the same time had helped to transform the cultural picture of the Hohokam area.

LEFT: *A Mimbres bowl holds the image of a rabbit with a starlike object attached to one foot. To the ancient people of the Mimbres Valley who made the bowl in the 11th century, the rabbit embodied the moon, much as it did among the Maya and other cultures of prehistoric America. In this unique example the juxtaposition of the "moon" to what appears to be a star with 23 points suggests the peoples of southwestern New Mexico witnessed the great supernova of A.D. 1054—the exploding star that created the Crab Nebula. In China, where astronomer Yang Wei-Te recorded the same event, the supernova was visible in daylight for 23 days. Seemingly confirming this coincidence, the rabbit and star on the Mimbres bowl mirror the general relationship between the moon and star as it would have been observed from both areas on the first day of visibility.*

ABOVE: *A modern descendant of the ancient Mimbres moon animal, a black-tailed jackrabbit (Lepus californicus) guards a cactus north of Benson, Arizona, in the San Pedro Valley.*

THE ANASAZI. In the language of the Navajo—who came relatively late to the Southwest from their homeland in the western Canadian subarctic—Anasazi means "ancient alien ones." Ancient they were but far from alien, as the Anasazi had deep roots in the Archaic desert cultures of the Southwest. Their hearth of development may be defined as that great sweep of scenic plateau land that reaches from the Grand Canyon eastward through Monument Valley to the Rio Grande, where their descendants—the Pueblos and the Hopi—live today. The rise, expansion, and fall of the Anasazi between about A.D. 600 and 1300 is regarded by many as *the* great cultural epic of the American Southwest.

The story of the Anasazi has gradually emerged from evidence uncovered at hundreds of sites. Among the most important are those that lie in two regions—Chaco Canyon, in northwestern New Mexico, and Mesa Verde, in extreme southwestern Colorado.

A typical early Anasazi settlement is Shabik'eshchee, or "Sun Picture Place," named for a petroglyph on its approach trail. Shabik'eshchee lies on the south rim of New Mexico's Chaco Canyon. There, in the 1920s, Frank H. H. Roberts of the Smithsonian Institution excavated 18 pit houses from the period between about A.D. 450 and 700, of which half were probably occupied at any one time, and a large kiva, one of the sunken,

The view from Hunt's Mesa reveals a stark northern Arizona landscape of desert dune and detritus punctuated by monolithic mesas rising to the sky. Now part of Monument Valley Tribal Park on the Navajo Indian

Reservation, this ancient land holds not only the remains of both Pueblo and Navajo ancestors but also the artificial boundaries of the "reservations" allocated to the modern Hopi and Navajo peoples of the region.

circular chambers that serve as hallmarks of most Anasazi sites.

A more recent survey of Shabik'eshchee by archaeologists W. H. Wills of the University of New Mexico and Thomas C. Windes of the National Park Service has increased the count of buried pit houses to 68 and recognizes the site as but one of 163 sites of the same period in Chaco Canyon and its environs. The kiva, one of the earliest known, may have served a ceremonial function similar to that of its later and modern counterparts. The site also revealed the adaptation of major innovations such as cotton cloth, the bow and arrow, and pottery—perhaps inspired by contemporary Mogollones—which eventually replaced baskets.

Between about 900 and 1150 the Anasazi population reached a peak across the entire Colorado Plateau, and spectacular clusters of dwellings rose in canyons and sheltered cliffsides—places such as Canyon de Chelly, Keet Seel, and Betatakin in northwest Arizona, and Mesa Verde in Colorado. The largest Anasazi constructions—the "great houses" such as Pueblo Bonito, Chetro Ketl, Pueblo del Arroyo, and Pueblo Alto lay in Chaco Canyon, a 12-mile valley bordered by parallel escarpments in the San Juan Basin of northwestern New Mexico.

ABOVE: *Panoramic view of the remains of the final construction stage of Pueblo Bonito shows the pattern of rooms, kivas, and other specialized spaces in the huge D-shaped complex. The distinct masonry style of each major period of building allowed archaeologists to reconstruct in detail the growth of the pueblo over time.*
RIGHT: *The setting of Pueblo Bonito near the north escarpment of Chaco Canyon provided both access to the valley floor and a defensible location.*

Of all the ruins in Chaco Canyon, Pueblo Bonito is the largest, even by Anasazi standards. A gigantic D-shaped structure, it was once five stories high around the curved portion, which backs up

to the north escarpment of the canyon, but a remnant of the fifth story was destroyed by a rock fall in 1941. The whole ruin covers about three acres of canyon floor and held around 650 rooms. Surprisingly, studies of these chambers by archaeologist Thomas C. Windes and others indicate that relatively few of them served as living quarters. The purpose of the rest remains unknown, although some scholars suggest that they may have served for storage. In addition, the great construction holds 38 kivas, including four of unusually large size, plus the largest, the centrally located Great Kiva, measuring an astonishing 65 feet in diameter. Tree-ring dates from the ruins show that the grand construction rose in stages between A.D. 850 and 1150.

Given the essentially nonresidential character of Pueblo Bonito itself, population estimates for it mean little, for it is possible that differing groups may have used the place periodically for various purposes. Whatever the case, the low count of actual living quarters yields a rough guess of fewer than 200 as a maximum number of inhabitants, with a calculation of about 2,000 for the entire canyon.

In the 11th century either emigrants from the core population in Chaco Canyon or awed imitators began to establish outlying settlements as distant as 100 miles. At least 150 of these are known across the region, each marked by a formal arrangement of rooms facing a plaza, and kivas—a plan seemingly modeled on the typical Chaco "great-house" scheme.

A remarkable system of roads at least partially linked many of these Chaco outliers to one another and to the core area of Chaco Canyon itself. Discovered through the analysis of aerial photographs, the roads, carefully scraped from the desert floor, run for a total, so far, of more than 400 miles over an area of nearly 60,000 square miles. Their makers clearly planned these roads to run straight and true—some even cross steep cliffs by means of footholds in the rock, when such areas could more easily have been skirted. As for the purpose of such a system by a culture without wheeled vehicles or beasts of burden, "we can only surmise," archaeologist Stuart Fiedel writes, "that the road network had symbolic significance, perhaps as a concrete manifestation of the unity of the Chacoan sociopolitical system."

The Chaco Phenomenon, as archaeologists sometimes call this unusual surge in the Anasazi archaeological record, may owe at least part of its prosperity to a thriving trade network that may have reached deep into Mesoamerica. Evidence from the Chaco area suggests contact with Mexican merchant groups who brought, for example, macaws—greatly desired for their feathers—in return for turquoise and other local commodities.

BELOW: *Necklace and ear pendants of polished turquoise came from a west interior room of Pueblo Bonito, where they had been carefully cached in antiquity, perhaps during a raid by outsiders. Archaeologist Neil M. Judd made the find. "It was so unexpected, so unforeseen," he later wrote. "A casual scrape of a trowel across the ash-strewn floor, a stroke as mechanical as a thousand other strokes made every day, exposed the long-hidden treasure."*

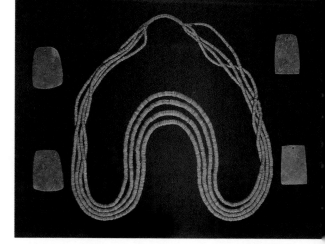

ABOVE: *A parade of polychrome effigy vessels from Paquime in northwest Mexico shows realistic renderings of hairstyles, clothing, and face paint.*

Physical anthropologist Christy Turner, II and the late Jacqueline Turner have reached an entirely different conclusion on the reasons for the Chaco Phenomenon—one also involving Mexicans as newcomers to Chaco—not as merchants, but as usurpers who instigated a two-century reign of terror and violence. Their evidence for this lies in a large number of butchered, broken, and burned human bones, often in deposits containing ten or more adults or children killed at the same time.

The pattern revealed by the overwhelming amount of evidence documented by the Turners suggests to them the precise kind of ritual cannibalism associated with war, violence, and human sacrifice as practiced by many Mesoamerican cultures of the time. Though many archaeologists see the Mexican connection as far-fetched, none can deny that such practices would have thoroughly terrorized and subjugated the Anasazi peoples of the Chacoan great houses.

Whatever the nature of things during the Chaco Phenomenon, nothing could counter the 50-year drought that began in 1130. The demands of the ever increasing population of Pueblo Bonito and the satellite towns had exhausted the land already, so even the planting schedule—set by the sun priests by

As the morning sun rises behind the sandstone pillar of Fajada Butte, an Anasazi sun priest greets it with a ritual offering of ground maize, crushed shell, and turquoise. From his special vantage point, he could forecast the

86

winter solstice exactly 16 days later, according to archaeoastronomers. Through such daily observations, priests maintained an accurate calendar for planting. Historic Pueblo practices provided the basis for this reconstruction.

their astronomical observations and formalized by ritual—didn't matter anymore. With this came the failure of trade and any prosperity that had held the Chacoan world together. By 1250 construction and routine use of the canyon had halted, and many if not most of its people had moved on, abandoning the large towns first, then the others.

Mesa Verde National Park in southwestern Colorado, the largest archaeological preserve in the United States, holds more than 4,000 separate sites, and hundreds more come to light each time wildfires scorch a new area. The Anasazi who lived in those sites inhabited a land far easier to farm than that around Chaco Canyon. More rainfall, plus the skilled use of terraces and check dams to conserve runoff and snowmelt, allowed the Mesa Verde farmers to make the best use of mesa tops, as well as their side canyons and ravines.

Like other Anasazi peoples, these had moved from the traditional pit houses to aboveground pueblos, each with small kivas, by A.D. 900. By 1200 they had created the large aggregated pueblos such as Cliff Palace and other famed examples tucked in hol-

A Colorado sunset briefly illuminates the masonry walls of Cliff Palace in Mesa Verde National Park. Builders of this extensive site, one of hundreds at Mesa Verde, situated it securely under the overhanging rim of the escarpment that defines the great "green table." Its square and round towers, rooms, and kivas served as living quarters, places for the storage of food and trade goods, and settings for religious rites.

lows below the tops of the escarpment walls surrounding Mesa Verde. These range in size from a few rooms, probably for a single family, to as many as 151 rooms and 23 kivas in Cliff Palace.

Many of these cliff dwellings, not only at Mesa Verde, but also at Keet Seel and Betatakin in the Kayenta area of northern Arizona, may have been constructed at least partly as defenses against other Anasazi pueblos, for by 1200 minor cycles of drought were already depleting the land—and discontent had grown hand-in-hand with the fear of starvation.

We now know more than we ever have about the nature of conflict among the late Anasazi, thanks to the efforts of Jonathan Haas and Winifred Creamer in northeastern Arizona, and years of investigation in the region of Zuni, New Mexico, by Steven A. LeBlanc and his colleagues.

Haas and Creamer chose the Kayenta area for their study because previous archaeological research had already established an accurate and precise chronology there, along with details of the history of its pottery styles and climatic changes. This information formed a solid basis for their study of the changes, mostly for the worse, that marked the last 150 years of the Anasazi occupation of the region.

From their thorough survey of ruins, rockshelters, and other remains in Long House Valley and other parts of the Kayenta area, Haas and Creamer were able to visualize related patterns of movement in which widely scattered settlements gradually evolved into close-knit clusters of cooperating communities. This suggested a basic change in social organization, whereby self-governing villages became parts of larger, more complicated political cooperatives engaged in the conservation of water and other essential resources. The increasing necessity to defend such settlements against marauding outsiders, Hass and Creamer believe, may well account for the presence of many late sites atop steep-sided mesas or cliff hollows. Such places not only could be easily defended, but also afforded vantage points for observation of the surrounding area—and for visual contact among allied community clusters.

Farther south, Steven LeBlanc found a similar situation among the Anasazi ruins of the Zuni River Valley, and for the same dreadful time period. For while relative prosperity had reigned during the relatively wet, warm climate between A.D. 900 and

1200 or so, the time afterward was one of stress, failed crops, malnutrition—and warfare.

In the El Morro Valley, LeBlanc and his colleagues had noted a repeating pattern of paired settlements in which the earlier one had been destroyed and burned and its building material salvaged to help build a more defensible settlement very close by. All such pairs dated within a five-year span, A.D. 1279 to 1284. Looking at the larger regional picture, it soon became apparent that these communities, much like their counterparts in the Kayenta area to the north, formed clusters of from 3 to 18 settlements.

"Each cluster," notes LeBlanc, "was spaced about 20 miles from other clusters. The empty spaces between the clusters contained good farmland. These open areas were uninhabited because they were indefensible 'no-man's-lands,' and too dangerous to live in."

The times apparently never improved for the Anasazi of Kayenta, Mesa Verde, and the El Morro Valley. By 1300 they all lay abandoned, their inhabitants departed to establish new lives on the upper Rio Grande and the Hopi mesas. A Zuni story, recorded just over a century ago, tells the story best:

"Fearing that never again would the waters refreshen their cañons, our ancients who dwelt in the cliffs fled away to the southward and eastward—all save those who had perished aforetime; they are dead in their homes in the cliff-towns, dried, like their cornstalks that died when the rain stopped long, long ago, when all things were new."

CHAPTER 5

THE SOUTHEAST

EPIC OF THE BUILDERS OF THE MOUNDS IN NORTH AMERICA'S EASTERN WOODLANDS

T HE STORY OF THE VARIOUS INDIAN GROUPS of southeastern North America is very much a story of different kinds of mounds or other earthworks they designed and built—and why— over the past 6,000 or so years. That amazing span reaches far back into the Archaic Period, to a time in the East when the balance between the human population and the available natural resources was—at certain times and in certain locales— weighted heavily in favor of a settled and prosperous life, even in the absence of agriculture.

As Stuart Struever's excavations at the Koster site in the Illinois Valley have suggested, a sedentary life was probably fairly common amid the rich and varied landscapes of the Archaic Period population of the East, between about 4000 and 1000 B.C. Recent investigations have revealed even more examples—mound sites that appear to be far more complex than the camps and seasonal settlements that archaeologists had come to expect for that time. This is why the academic community was unprepared in the early 1960s when Sherwood Gagliano reported radiocarbon dates at the older end of that range for a cluster of mound sites in south eastern Louisiana. Now, thanks to the efforts of Michael Russo, Joe Saunders, and other investigators working in northeastern Louisiana, we know of more such places; their dates—between 4000 and 2000 B.C.—make them the earliest mounds we know of on the North American landscape.

Watson Brake consists of 11 mounds built in a rough oval with its long axis measuring about 975 feet, with all but four mounds connected by a ridge. The highest mound rises about 25 feet. Archaeologically, little is yet known about the place except for some Archaic stone points, tons of fire-cracked rock (perhaps the remnants of cooking activity), and thousands of fish bones, remains of shellfish and snails, and plant remains, along with a few "Watson Brake objects"—fired clay cubes or cylinders of unknown use.

All together, nine mound sites comparable in age to Watson Brake lie scattered in eastern Louisiana. Their very presence suggests something new on the landscape of Archaic Period Louisiana. As archaeologist Jon Gibson, who probably knows the prehistory of the region as well as anyone, notes, "Mounds are conspicuous, permanent reminders of a group's labor and existence. Whatever

PRECEDING PAGES:
Sunrise reveals wetlands and woodlands of northern Ohio. Such an environment spawned some of ancient America's most important and interesting cultures, among them those of the Adena and Hopewell peoples of the Ohio River Valley, who together endured for more than a thousand years.

OPPOSITE: Human effigy pipe of stone was discovered during excavation of the Adena Mound, Ross County, Ohio, in 1901. The eight-inch-tall carving depicts a man wearing a loincloth, headdress, and elaborate ear ornaments—and apparently afflicted by dwarfism, a goiter, and rickets. The mouthpiece of the smoking tube is at the top.

rituals might have been enacted on or around them, whatever purposes they may have served, and whatever their raison d'être, mounds can be understood as symbols of a community's identity and capacity for common action."

Jon Gibson has devoted much of his career to unraveling the secrets of another mound site in the region. Later than Watson Brake and its enigmatic contemporaries, Poverty Point is a name instantly recognizable to any student of the American past. Dean Snow, one of the best in the business of summarizing the ebb and flow of ancient American culture history, characterizes the Louisiana site as "perhaps the most impressive Archaic site in North America."

Though erosion has taken its toll, six concentric ridges, interrupted by radial alleyways, still define the heart of Poverty Point. In plan, these ridges form a partial oval three-quarters of a mile across and open to the east, where it ends at the edge of the Bayou Macon—and the alluvial plain of the Mississippi. On the outer edge of this remarkable series of earthworks rises the huge irregular mass of Mound A that appears to some as a gigantic bird effigy. Radiocarbon readings help date this core area of Poverty Point to around 1100 B.C. Five other mounds are situated near the ridges or in the vicinity. Fifteen other mound sites, related to Poverty Point by virtue of their dates and content, lie in the same general region, some within a 25-mile distance up and down along the terrace that follows the Bayou Macon.

Deep deposits of refuse excavated between the ridges at Poverty Point resemble those of a large town, and Jon Gibson and others who have worked at the site estimate that some 600 houses could have stood there. In addition, more than 30 satellite sites, scattered up and down the Mississippi Valley for several hundred miles, all feature to some degree artifacts and art similar to those at Poverty Point. These include fine lapidary work, carvings of bird pendants in semiprecious stone and other traded materials, early clay figurines, and the ubiquitous "Poverty Point objects"—small baked-clay cylinders, balls, bi-cones, and other shapes, often decorated by grooving, punching, or squeezing. Most investigators see these as objects for heating water or for use in baking pits.

OPPOSITE: *The distinctive type of notched stone projectile point shown here occurs in association with Watson Brake, Frenchman's Bend, and other early mound sites in eastern Louisiana and helps date those places to the Archaic Period, between 4,000 and 6,000 years ago.*
RIGHT: *The concentric earthworks at Poverty Point, in the same region, distinguished in this NASA satellite image by plantings of red clover, dates from around 3,000 years ago, and stands as the earliest such system of complex earthworks yet known in eastern North America.*

There is still much to learn about Poverty Point and its neighbors, and the prospects look exciting for Gibson and his colleagues. "I can see the seeds of social differentiation and raw economic materialism sprout from the mound tops," he writes, "regardless of the manner of ritual that sowed them. One of the things that grew out of these Middle and Late Archaic mound-building contexts in the Lower Mississippi region was a far-reaching exchange system, which is often accorded culture status—Poverty Point Culture."

IN THE MEANTIME, OTHER MOUNDS OF EARTH were beginning to rise in what is now the Ohio Valley, from Indiana to West Virginia, with

The Great Serpent Mound of Adams County, Ohio, lies as if uncoiling atop a rocky ridge overlooking Brush Creek. About four feet in average height, the body of the earthen snake extends for

some 1,300 feet, from the spiral whorls of its tail to its open mouth, the latter poised as if to devour the oval of earth just beyond. Recent investigations have dated it to the early 11th century A.D.

some as far eastward as the Chesapeake Bay region. The people who built these monumental memorials to their honored dead in the centuries after about 600 B.C. are known to archaeologists as the Adena culture, named for the great mound excavated in 1901 on the Adena estate near Chillicothe, Ohio. Their span of existence also defines the early part of what archaeologists term the Woodland Period of eastern North America, 1000 B.C. to A.D. 800.

The central theme of Adena culture lies in its special concern for the dead, for each of the conical burial mounds conceals one or more burials, with the most important interred in elaborate crypts built of heavy logs. The Cresap mound in West Virginia's northern panhandle held a succession of 54 burials that spanned much of Adena culture history. In downtown Moundsville, not far away, the Adena created their largest known burial memorial, the Grave Creek Mound, 240 feet in diameter and originally 70 feet in height—roughly equal to that of an eight-story building.

Archaeologists can only speculate on the ceremonies that accompanied the Adena elite to the tomb, along with hammered copper jewelry, tubular stone pipes, and axes and ornaments of polished stone. Strong evidence of Adena shamanism shows up as well. The remains of an elk-antler headdress lay near one of the Cresap burials, and a toothy section of bone carefully cut from the mouth of a wolf came from the Niles-Wolford Mound in Ohio. Another burial, from the Ayers Mound in Kentucky, revealed the purpose of the wolf bone. In what was certainly a shaman's burial, a thinned wolf palette had been placed in the individual's mouth, in the space made by the removal of the upper front teeth. The shaman's bone had healed thoroughly, indicating the dentistry had been done during life. These rare samples make it easy to imagine the Adena shaman in the ritual role of animal impersonator, an effect perhaps enhanced by wolf-skin mask and robe.

Small stone tablets, polished and bearing bird motifs and curvilinear abstractions, best show the high achievement of the Adena artist. Such tablets may have functioned as stamps for body or textile paint, for some still contain red pigment in their deep recesses.

ADENA CULTURE BEGINS TO VANISH from the archaeological record beginning around A.D. 100 and seems to have been gradually replaced by Hopewell, another Woodland culture, in the same

general area of the Ohio Valley. No one yet knows the circumstances of the changeover. Some suggest an influx of early Hopewell into the present general area of Ohio from Illinois. Others see a simple cultural evolution of Adena into Hopewell. Whatever the nature of the event, the Hopewell way emerged decisively after 200 B.C.

Olaf Prufer of Kent State University, the dean of Hopewellian studies, sees the pattern of Hopewell settlement as one of dispersed hamlets near the great earthworks that serve as the main hallmark of Hopewell culture. Like their Adena predecessors, the Hopewell people lived by intensive harvesting of wild plants and by hunting. They also cultivated phalaris, polygonum, chenopodium, and perhaps sunflower, along with some cucurbits and other native plants. Drawing much of their art style,

Artist's reconstruction of a Hopewell burial feast is based on clothing and ornaments depicted on figurines or artifacts recovered from excavations at various sites in the Ohio Valley—and on educated speculation. In this scene a shaman, clad

in a feather mantle, shell necklaces, copper ornaments, painted cloth, and a headdress of copper antlers adorned with pearls, makes a ceremonial offering, while dancers wait to participate.

death cult, and mound-building practices from the Adena, the Hopewell elaborated them into what stands as the cultural zenith of the Woodland Period, and nowhere is this more obvious than in the sheer size and scope of the grand system of earthworks, most of them now obliterated, that once covered most of the landscape in and around present-day Newark, Ohio.

Work by Bradley Lepper of the Ohio Historical Society in searching out old maps of the Newark earthworks has paid off handsomely by permitting him to reconstruct the amazing system as it was originally built.

It begins on the west with raised ridges of earth forming a 20-acre perfect circle linked by parallel embankments to a nearby 50-acre octagon containing eight mounds. A mile to the southeast, a 30-acre circle or ridge and ditch is open on the north end, where more parallel embankments connect it to a 20-acre square, more complicated alleyways, and a gigantic oval of some 50 acres. The oval, in turn, held 11 burial mounds—one originally 140 feet long and 15 feet high. The west and east systems are themselves linked by two grand passages enclosed by embankments, and a third such roadway goes off the map to the south. And amid all these main works appear dozens upon dozens of small mounds, circles, and ditches.

Where did the south roadway lead? By means of sporadic surface traces faintly visible in old aerial photographs of the region, Lepper has tentatively plotted the path of what he has dubbed the "Great Hopewell Road," a perfectly straight avenue, probably enclosed by low parallel ridges, that appears to have once linked Newark with still another complex of earthworks in Chillicothe, nearly 60 miles away! "When I began my investigation of the Great Hopewell Road," Brad Lepper recalls, "one prominent Midwestern archaeologist advised me that the Ohio Hopewell simply could not have built such a structure because they lacked the population, the subsistence base, and the political organization requisite for undertaking such long term public works projects....I submit that this is the fundamental mystery of the Hopewell: How were they able to achieve what our theories suggest was beyond their capabilities? I further submit that our theories are inadequate if they do not take into account the possibility that the Ohio Hopewell were building great roadways connecting widely separated centers of social and religious activity more than five centuries before the Anasazi erected their first pueblo."

Many archaeologists attribute the Hopewell achievement to

LEFT: *A hand nearly one foot high, meticulously cut from sheet mica traded from the southern Appalachians, exemplifies one class of Hopewell art—probably made for ritual display and adornment—that includes silhouettes of animal claws, decapitated human figures, and animal forms.*
OPPOSITE: *Polished stone "platform pipes," such as the example showing a roseate spoonbill perched atop a fish, typify another major class of Hopewell art. Often inlaid with pearl eyes or shell teeth, these delicately carved pipes depict birds, mammals, fish, and other creatures of the area.*

56750

a prosperity that came in part through a trade network that covered much of central North America and provided exotic raw materials for the artists and craftspeople of ancient Ohio. From Yellowstone, for example, came black obsidian that was chipped into huge, thin ceremonial blades; from the Upper Peninsula of Lake Superior came copper nuggets to be hammered into embossed breastplates, ear ornaments, or ritual weapons; large sheets of mica from the North Carolina mountains were cut somehow into glistening silhouettes of hands, bird claws, animals, and headless humans; from the Rocky Mountains arrived grizzly bear teeth; and from the Gulf Coast came shells for the manufacture of bowls and ornaments.

Lepper sees the picture of Hopewell economy somewhat differently. "There is considerably more evidence of exotic material coming into Ohio," he notes, "than of Ohio goods going out. No one has explained satisfactorily how this disparity makes sense in terms of the economics of trade." In conclusion, Lepper sees Newark and other great earthwork complexes as sacred pilgrimage destinations. "Pilgrims from across Ohio and eastern North America would have to come to places such as Newark and Chillicothe with offerings of rare and precious items. Perhaps these offerings were in payment for the healing of an illness, for the counsel of an oracle, or for the acquisition of spiritual power."

Whatever the nature of the Hopewell interaction network, it doubtless helped carry the unmistakable stamp of the Hopewell cult and culture to many developing cultures of the Southeast. The Marksville site in central Louisiana resembles those in Ohio, and its pottery imitates the incised Hopewell wares. From the Mandeville and Kolomoki sites in Georgia all the way to Fort Center in south-central Florida, researchers have found Hopewell-like figurines and ornaments, along with implements

Platform mounds, 20 in all, surround the great plaza at Moundville, Alabama, on a terrace overlooking the Black Warrior River. Moundville flourished between about A.D 1200 and 1500 as the center of a powerful Mississippian

chiefdom. Pottery, stone paint palettes, and other artifacts associated with the Moundville elite often bear skulls, crossbones, rattlesnakes, and other symbols and motifs of the Southern Cult, a widespread religious movement of the time.

chipped from the distinctive rainbow-colored stone from Flint Ridge in southern Ohio.

AFTER ABOUT A.D. 400, the Hopewell way began to fade in the area of its greatest accomplishments and also took on different forms in the Southeast. A new pottery type, involving bird and animal images on openwork vessels marked by curvilinear openings and incising, came into style at Weeden Island, Florida, and Kolomoki. Most importantly, many sites, including Swift Creek and Kolomoki, both in Georgia, and a few other Late Woodland sites of around A.D. 700, hold a different kind of mound—mounds with flat tops. From this time on, the innovation of the platform mound would not only serve as the main manifestation of ancient earthen architecture in the Southeast; it also helped to define the Mississippian Period.

Aside from mounds, archaeologists define Mississippian culture—so named because it may have developed along that great river—by the presence of large villages, many with open plazas;

The rising sun of the autumn equinox aligns itself with a wooden post inside a reconstructed sun circle, or "Woodhenge," at the great Cahokia site in southern Illinois. The circle, 410 feet in diameter and one of several built over time at the Mississippian center, may have helped establish a solar calendar for the planting of corn and for other purposes.

intensive maize agriculture; and, in some places, art and iconography of extraordinary richness. To many archaeologists and art historians, the platform mounds, carefully arranged around plazas, the sophisticated shell-tempered pottery in a variety of painted and effigy styles, and some aspects of the art style demonstrate clear influences, perhaps indirect, from Mesoamerica—as do the strains of maize and beans present at the sites. So far, though, no single artifact of undisputed Mesoamerican manufacture has ever been found in the Southeast. If such a specimen ever comes to light, one archaeologist told me only half in jest, it will probably be at Cahokia—the largest of all Mississippian sites.

Cahokia lies eight miles east of St. Louis, on the Illinois side of the Mississippi River. Monks Mound, named for a colony of Trappists who lived nearby in the early 1800s, dominates the site. The massive construction covers more than 14 acres and rises in four terraces to a height of 100 feet above the great alluvial plain known as the American Bottom. Monks Mound is the largest single prehistoric construction on the continent north of the ruins of Teotihuacan, Mexico. Cahokia is situated near the confluence of three major rivers, the Mississippi, the Missouri, and the Illinois, and embraces four different ecozones—a situation similar to the setting of the Koster site, some 60 miles to the north, a settlement that successfully endured from Archaic times on.

Much of Cahokia's former grandeur has been obliterated by modern housing, shopping centers, and highways. Between A.D. 1100 and 1200, the peak of its prosperity, Monks Mound faced a great plaza that marked the heart of a Mississippian city estimated to have held somewhere between 8,000 and 20,000 people (the wide range of the figure stems from lack of evidence either on the precise extent of the city or the complexity of its organization). Directly across the plaza rose another large truncated mound with a conical mound beside it. Another 120 or so mounds in various parts of the city held official temples, residences, or burials. Houses—their walls covered with mats of cattail or bulrush, their roofs of thatch—filled most of the area within the 12-foot-high palisade of logs that surrounded the great rectangle of the city center. Beyond the great wall lay still more houses and more mounds, along with fields of corn, pumpkin, sunflowers, maygrass, and marshelder.

Beginning in 1961, archaeologists at Cahokia revealed several perfect circles of postholes—round dark stains on the excavation floor. The largest of these "American Woodhenges," as they were dubbed, measured 476 feet in diameter. Analysis showed that these postholes once held huge posts of red cedar, with a larger pole marking the center. The five known woodhenges, built over time in the same general spot, appear to have served as markers of the solstices and equinoxes, the latter as times for planting and harvest.

By A.D. 1250 Mississippian settlements similar to Cahokia, but smaller, occupied most of the great river valleys of the Southeast,

LEFT: *The remarkable Birger figurine, an eight-inch carving of bauxite, came to light at the BBB Motor Site near Cahokia, Illinois. It depicts a female kneeling on a great supernatural serpent whose teeth are those of a puma or other feline carnivore. The woman's left hand rests near the serpent's head, while her right hand wields a hafted stone hoe near the monster's tail, which forks and forms gourd plants, the whole suggesting the theme of agricultural fertility.* OPPOSITE: *One of eight embossed copper plates, each around a foot high, discovered when a Missouri field was plowed in 1906, shows a man in the form of a peregrine falcon, a common icon of Southern Cult art. He is adorned with a beaded forelock, ear ornaments, and a complex headdress featuring a jawless head duplicating the "weeping eye" of the main figure.*

up and down their namesake river from Wisconsin to Mississippi, and from eastern Oklahoma to central North Carolina.

Along with an elite group of priest-shamans and warriors, powerful chiefs, as heads of their clans and lineages, governed the larger Mississippian centers such as Cahokia; the Powers Phase towns of southeast Missouri; Moundville, Alabama; Ocmulgee, Georgia; and Cofitachequi, South Carolina. Many of these places, dominated by central mounds and plazas, became important regional capitals in a sea of major and minor chiefdoms that formed a loose patchwork over much of the Southeast—and whose fortunes rose and fell with changing patterns of climate, warfare, and trade.

Objects accompanying burials in Cahokia Mound 72 and elsewhere only hint at the pomp and circumstance attending members of the Mississippian elite—cedarpole litters, smoothly ground discoidal game stones, and bundles of arrows tipped with delicate triangular points of exotic cherts.

SOMETIME IN THE 14TH CENTURY—some would put it even earlier—many of the Mississippian centers were caught up in what appears as a powerful religious movement related to war, death, and a special world of supernatural beings—winged spiders, winged and horned snakes, antlered fish, and one particularly complex figure, a cat-like creature with the head of a man and the tail of a snake. Archaeologists tentatively call this movement the Southern Cult and identify three Mississippian sites as its main known centers—Etowah, Georgia; Moundville, Alabama; and Spiro, in the Caddoan-speaking country of eastern Oklahoma—but remains of the cult occur from the northern Mississippi Valley to Florida.

In the fall of 1933 the Craig Mound, a tall conical hump of earth with three smaller mounds connected to it, stood with other mounds in a cultivated field beside the Arkansas River, near Spiro, Oklahoma. That November, six unemployed local men—victims of the Great Depression—formed what they called the "Pocola Mining Company," signed a two-year lease with the property owner, and began a commercial venture that, in retrospect, ranks

as perhaps the single greatest tragedy in the history of American archaeology. Deaf to the pleadings of archaeologists and local citizens, the diggers pitted and tunneled the Spiro mounds and trundled out in wheelbarrows what witnesses would later term "one of the most amazing caches of ceremonial material ever found in the mound area"—and sold most objects on the spot. By the fall of 1935, when the lease ran out and the members of the Pocola Mining Company had sunk into mutual dissension and disagreement over the spoils, they reportedly dynamited their crude tunnel and departed the remains of the Craig Mound.

Gossip and anecdote suggest that the looters of Spiro encountered an intact mortuary chamber in the center of the Craig Mound that held a vast array of material—an astonishing loot that included 120 or more pipes, 23 of them in the form of humans or animals; 11 human figures and 40 masks of cedar; 50 delicately chipped maces or swords; and nearly 200 huge conch-shell bowls engraved with intricate scenes and cult motifs—some perhaps depicting an early form of the Green Corn ceremony, or *Busk*, a ritual of renewal still practiced by the modern Muskogee and Cherokee.

ABOVE: *A large mound at the Etowah site in northwest Georgia served as a burial place for nobility. Now restored after centuries of flooding by the Etowah River and decades of cultivation and excavation, the mound likely supported a structure that held the honored dead until the time came for burial.* OPPOSITE: *A counterpart in function to the long-gone Etowah mortuary temple is depicted in a watercolor by John White, who saw the wooden building near the coast of present-day North Carolina in 1585.*

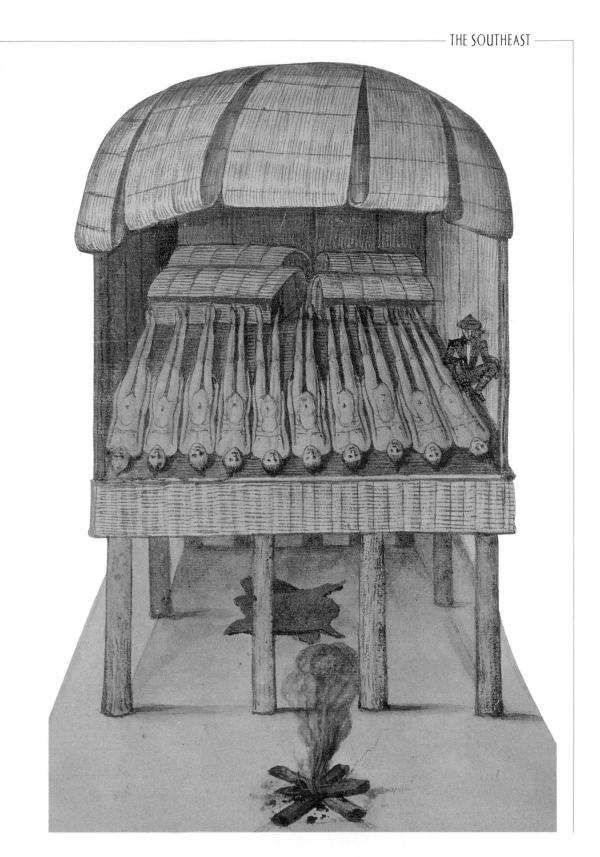

Among the most intriguing of the objects and materials ripped from the bowels of the Craig Mound are the remains of textiles—fragments of mantles made of polychrome cloth, rabbit hair, and feathers, and women's skirts of the same special materials. All these pieces of weaving serve as mute but eloquent testimony to the vast amounts of such material that have been forever lost through the natural process of decay over time. At Spiro these amazing specimens survived even the looters.

At the Etowah site in northwest Georgia the burials and tombs excavated by Lewis Larson from Mound C in the 1950s provided a look at the regalia and motifs associated with the Southern Cult. The burials contained large plates of embossed copper depicting peregrine falcons and other cult icons. One such plate, found at Etowah in the 1930s, shows a man in the guise of a falcon, wearing a bead-and-conch-shell necklace, wielding a stone sword, and holding a rattle in the form of a jawless head. Larson found all these artifacts, represented by the real objects—the shell necklace, 30-some-inch blades of Tennessee flint, and a copper-covered wooden effigy rattle—in other Mound C tombs.

Male and female figures of north Georgia marble probably portray two elite individuals of ancient Etowah, Georgia. The painted effigies—the taller is nearly two feet high—were set inside the sacred confines of the mortuary temple near the corpses of the pair until their disarticulated skeletons were removed from the building and interred, along with the statues and other objects of high status, in a large carefully prepared, log-lined tomb in the edge of a mound.

Burial 20, which I helped Larson excavate in June 1954, has always held a special place among my own recollections. The five-by-fifteen-foot tomb, enclosed by a wall of vertical logs, held the disarticulated skeletons of several individuals of high rank—and, in one end, where they had been dropped into the crypt, two seated statues of polished and painted Georgia marble, one male, one female. Each is about two feet high. These probably represented two of the individuals in the tomb—elite members of Etowah society who had lain in a mortuary house, accompanied by their portrait statues, until the flesh had fallen from their bones. Eventually their disarticulated remains—and the statues—had been ceremoniously relegated to the earth.

The peak of the Southern Cult—and the Mississippian mound centers themselves—appears to have passed by the early 16th century, shortly before the appearance of Hernando de Soto and other European explorers. By the time of European settlement the bright flame had vanished from the once-proud domains of Cahokia, Etowah, Moundville, and the others—and they lay abandoned and overgrown, vulnerable to all the future would offer them.

CHAPTER 6

MESOAMERICA

THE RISE AND FALL
OF CITIES AND CIVILIZATIONS
IN ANCIENT MEXICO
AND CENTRAL AMERICA

MESOAMERICA is a land defined not by modern political boundaries but rather by the ways of its inhabitants over the past four or five millennia. Anthropologists and archaeologists who have studied the area point to a unique combination of cultural traits, among them maize agriculture; a complex religious system featuring a large pantheon of gods governing everything from nature to warfare; a ritual ball game played in specially constructed stone courts; the construction of large public buildings, including the so-called pyramids—really stepped platforms—of stone; extraordinary skill in the arts and astronomy; the use of a complicated calendar system; and, in some cases, the use of hieroglyphic writing, which we see on everything from stone monuments to screenfold books of paper or deerskin.

The Mesoamerican landscape is one of extraordinary contrast. It begins in present-day north-central Mexico as a broad desert plateau flanked by mountain ridges and narrow coastal plains and ends in the forested mountains of western Honduras and El Salvador. Its total area is less than five times the size of Texas, yet Mesoamerica embraces nearly all the extremes of topography and climate to be found anywhere on Earth. Near Mexico City, glistening snow-capped volcanoes form the eastern border of the Basin of Mexico, a lofty, relatively arid plateau that defines the heart of the Mexican Highlands. To the east, some 800 miles distant, lies the flat, forested plain of the northern Yucatán Peninsula, home of the lowland Maya. Both of these extremes—and many in between—lie in the tropics, with alternating seasons of rain and drought.

Because culture provides the essential definition of Mesoamerica, the boundaries—and shape—of the area changed through time, as did its cultural attributes. By about 2000 B.C. Mesoamerica was a world of farmers, and the people of this world had just begun to forge the distinctive Mesoamerican pattern of culture that would endure for more than 3,500 years— and continue in some ways to the present day. More for convenience than as accurate tags of cultural progress, archaeologists divide the long history of the ancient Mesoamericans into major spans of time. These include the Preclassic Period, from about 2000 B.C. until around A.D. 250; the Classic Period, from 250 until around 900; and the Postclassic Period, from 900 until the coming of the Spaniards in the early 16th century.

The story of ancient Mesoamerica is one of many different

PRECEDING PAGES: *At the heart of the ancient Maya ruin of Copán, Honduras, a spider monkey strides by a stairway of the stepped platform, or "pyramid," that defines the center of the city's Great Plaza. With their four staircases, such structures served as metaphors for the four-cornered universe of Maya belief.*

OPPOSITE: *A colossal head from San Lorenzo, Veracruz, Mexico, probably portrays a powerful Olmec chieftain who held sway over the regional capital some 3,000 years ago. Such mammoth portraits—the largest weighs more than 30 tons—were carved of basalt transported some distance from the Tuxtla Mountains. To date, 17 colossal heads have been found at San Lorenzo, La Venta, and other Olmec sites in the same Gulf Coast region.*

The overgrown archaeological site of Laguna de los Cerros reflects the regular planning that is a feature of most major urban
and religious centers of ancient Mesoamerica. A common Olmec scheme is evident in at least part of the site, in the two

large earthen pyramids on a north-south line, flanking a walled plaza. Looters' holes show up as circular depressions in many of the mounds. Scientific excavation nearby has uncovered a workshop where basalt statues and stelae were carved.

121

peoples whose individual dramas often overlapped on the vast and varied stage of the landscape. Those peoples included the Preclassic Olmec of Mexico's Gulf Coast lowlands; the Classic Period city-dwellers of the gigantic metropolis of Teotihuacan on Mexico's high central plateau; the Zapotec and Mixtec peoples of the mountains and valleys of Oaxaca to the south; and, far to the east, the Maya of the Yucatán Peninsula. Later, during the Postclassic Period, the enigmatic, ill-defined, and legendary Toltec peoples would fill the gap left by the mysterious demise of mighty Teotihuacan. Finally, the opportunistic and militant Aztec, or Mexica—who would eventually provide the modern nation with their name and the central device for their national flag—entered the Basin of Mexico, welded a powerful tribute empire over much of Mesoamerica, and created Tenochtitlan, their island capital, whose splendor would so awe Cortés and his soldiers in the late autumn of 1519.

Olmec and other farmers of the Mesoamerican lowlands took full advantage of a humid, tropical climate and its rivers, wetlands, and forests such as this one, site of a nesting snowy egret. In the tropical dry season, both lowland farmers and inhabitants of more arid zones of the highlands employed irrigation or raised fields laced with canals to ensure the production of food.

AN IMPORTANT BEGINNING of the Mesoamerican story took place east of Mexico City, about halfway to Yucatán, in the humid tropical lowlands of Veracruz and Tabasco states. The locale is marked by the forested slopes of the Tuxtla Mountains, a compact range of volcanic basalt that rises beside the coast of the Gulf of Mexico, and it was near here, in 1858, that the first remarkable Olmec archaeological discovery took place. An Indian laborer clearing jungle growth near the settlement of Tres Zapotes, Veracruz, saw what he thought was a huge kettle buried bottom up in a clearing. Hurried excavation revealed an enormous human head with stolid features and a helmet-like cap, carved out of basalt.

While the Tres Zapotes head languished at its discovery site, unpublished except for a brief speculative account of its possible "Ethiopian" origin, other heads came to light in 1925 at nearby La Venta. These, along with strange and exquisite jade carvings said to be from the same region, led Smithsonian Institution archaeologist Matthew W. Stirling into the land of the Olmec in 1938. By 1945, with the help of the National Geographic Society, Stirling had completed a total of six major expeditions—and the revelation of what he called "a culture hitherto unsuspected by the public."

Before Stirling's work many thought that the colossal heads, the jades, and other finds in the swamps and jungles of the Gulf Coast area belonged to the "Olmec," or "rubber people" described by the 16th-century Aztec as those who lived in a land of abundant rubber trees. After working at the major sites of Tres Zapotes, Cerro de las Mesas, La Venta, and—largest of all—San Lorenzo, Stirling maintained that these were not remains of the Postclassic Olmec, but much, much earlier. He was right, but the name stuck. We now know that Olmec culture flourished between about 1200 and 400 B.C., and that at least two major centers served in succession as regional capitals under powerful rulers—San Lorenzo and La Venta.

The plateau that holds the archaeological site of San Lorenzo rises beside the meandering Coatzacoalcos River in an area of swamps and streams some 35 miles from the Gulf Coast. From the air the plateau, its edge cut by deep erosion gullies, seems to take the shape of a great bird in flight—an effect that, combined with the enormous and deep areas of fill added to the feature in antiquity, once led to the thought that perhaps the whole plateau was artificial.

From the road that approaches the site and the adjacent town of San Lorenzo Tenochtitlan (no kin to the Aztec capital), the plateau makes a long, almost imperceptible, bulge, a dark band,

on the horizon as one approaches it on the road from Acayucan. In 1945, when Matthew and Marion Stirling, along with National Geographic photographer Richard Stewart, made the same journey by horseback, they found five colossal heads, most of them half-buried in the ravines where they had been revealed by erosion. Later, archaeologists Michael D. Coe and Richard Diehl discovered that the inhabitants of San Lorenzo had intentionally buried many of their largest sculptures, including a statue of a spider and various beings best described as were-jaguars—half human, half feline—on specially prepared floors.

In 1994 archaeologist Ann Cyphers of the National Autonomous University of Mexico unearthed another head at San Lorenzo, the tenth now known from the site, and the seventeenth in all. Matthew Stirling's early opinion that these heads portrayed the prominent leaders of the Olmec centers where they were found indeed appears to have been their function. At any rate, each is without much doubt the portrait of an individual—one powerful enough to compel subjects to go as far as 50 miles to bring gigantic boulders of basalt from the Tuxtla Mountains, and then to commission sculptors to render the portrait heads.

Both Ann Cyphers and Rebecca González, who headed recent excavations at La Venta, are learning much about Olmec life at the two places. Although both sites were occupied by the beginning of Preclassic times, they flourished at different times—San Lorenzo between about 1200 and 850 B.C.; La Venta from about 800 until around 400 B.C. Both places in their heyday probably matched Rebecca's succinct description of La Venta: "a prosperous community of fishers, farmers, traders, and specialists, such as the artisans and the sculptors."

The foundation of Olmec achievement lay with the accomplishment of the farmers who sustained this remarkable culture. Indeed, most of daily household activity must have revolved around the preparation of maize and other cultivated products such as beans, squash, chili peppers, and manioc. Hunting, fish-

ABOVE: *Four-inch Olmec reclining figurine of polished stone may represent a dancer or an individual in a shamanistic trance. Even its intended orientation is puzzling, for it can be suspended equally well in a vertical position.*
OPPOSITE: *The eight-inch-tall Olmec figure of polished serpentine rubbed with red pigment represents a shaman, perhaps also a ruler, caught in the transformation between human and jaguar forms.*

ing, and turtle collecting provided much of the protein, as they do for the present-day inhabitants of the region.

The Olmec were ruled by powerful kings. Sadly, there are but a handful of rudimentary hieroglyphs carved during this period, so we have no names to link to the inscrutable faces of stone that glower from beneath the ball-game helmets that adorn the colossal heads. Clearly meant to inspire awe, these great stones—some weigh as much as 30 tons—seem to have been set near points where trails could have entered the towns.

Such trails reached the other way as well, from the Olmec Gulf Coast "heartland" to virtually every corner of Mesoamerica, for the Olmec elite demanded luxury goods of exotic materials, including jade and obsidian, to help maintain their hereditary status. For this reason we find traces of the Olmec from the highlands to the most distant lowlands. At Teopantecuanitlán, in Guerrero state, Mexico, fierce Olmec were-jaguars glare from a ritual patio wall, while Olmec paintings decorate the darkest and deepest recesses of caves in the same region. At the great highland site of Chalcatzingo, Olmec bas-reliefs of plants, rain, a

Armless two-foot-high wooden busts from the Olmec site of El Manatí Veracruz, Mexico, date to some 3,000 years ago. Their extraordinary preservation, along with ceremonial rubber balls and other artifacts of wood and other perishable materials, came about from the unique nature of the site of their discovery—a muddy series of springs at the base of a hill that served as a ceremonial depository for sacrificial offerings.

sacred cave, and other figures decorate the cliff walls. At distant Copán, Honduras, Olmec jades and ceramics occupy the earliest layer of the site—later a capital of the Classic Maya. On the Pacific slope of Guatemala many remains as old as those in the distant heartland seem similar to the Olmec art style as well, but the archaeological jury is still out.

Along such trails passed another kind of commodity—ideas that we see as pervasive in Preclassic times—notions of the jaguar, the shark, and other creatures as companion spirits and, sometimes, as symbols of great power; and the idea of the sky, the Earth, and the netherworld of caves and water as the main parts of a powerful and capricious universe.

There are many things that we may never know about the Olmec—their language, their songs, their chants, their great ceremonies, their myths, and their textile arts. What we do know, however, is that the Olmec at their peak of cultural achievement had organized a system of life, religion, and ritual that worked. Not only did it serve the Preclassic world of Mesoamerica, but it also set the pattern and style, indeed, the very foundation, for all that took place over the next 2,500 years. The only part of the Mesoamerican pattern to which the Olmec did not contribute directly was the building of true cities, although the sheer size of San Lorenzo suggests to Ann Cyphers and others that it may have functioned as an urban center.

The first recognizable achievement of a true Mesoamerican metropolis became manifest some six centuries after the end of La Venta, and far away, in a side valley of the Basin of Mexico. There, high on the central Mexican plateau, there grew a city so large, so imposing, and so powerful that it became a legend even in its own time—between about A.D. 200 and 700 or, in terms of Mesoamerican archaeology, most of the Classic Period, which the city helps to define.

TEOTIHUACAN. The very name, Aztec for "Abode of the Gods," inspires awe for those who have never been there. For those who have actually stood at its center and witnessed what the creators of the great city conceived of as the crossroads of the cosmos, even the grandiose name seems inadequate. Even in its present state of a ruin reduced by time and the encroachment of modern

Its sheer mass dwarfing a throng of visitors, the Pyramid of the Sun dominates the main axis—the so-called Street of the Dead—and the distant Pyramid of the Moon at the heart of the ruins of Teotihuacan, largest city

ever built in ancient America. Teotihuacan flourished between about A.D. 100 and 650, and through trade and other means dominated the politics and economy of much of Mesoamerica during the Classic Period.

development, the reality of Teotihuacan is indescribably over-powering. Even visitors who have seen most of Mesoamerica's great ruins, from Tula to Tikal, are often amazed by their first visit to the Abode of the Gods. At the height of its prosperity Teotihuacan held as many as 200,000 inhabitants, rivaling Shakespeare's London of a thousand years later. The city outlasted its contemporary, imperial Rome—and this in the setting of a high and arid upland, hardly an ideal place for maize farming.

The city was carefully situated on what must have been an extraordinarily significant spot of earth defined by a fortuitous combination of sacred mountain peaks and an underlying system of holy caves—and it was carefully planned to harmonize with all this sacred geography. Based on a gigantic cross laid out in accordance with survey marks—the crosses and circles pecked into rock outcrops on surrounding viewing points still exist—the orientation of the main axis of the city was approximately 15 degrees east of north. This put the axis—the great ceremonial way that the Aztec later called the "Street of the Dead"—in line with Cerro Gordo, the principal peak to the north. Thus the Pyramid of the Moon, at the northern end of this axis, serves as an artificial counterpart to the mountain behind it.

The line perpendicular to this north-south axis was established about a mile down the Street of the Dead and adjusted very slightly to line up with a point on the western horizon where the Pleiades—a cluster of seven stars linked to the Mesoamerican calendar— set at the time of the founding of Teotihuacan.

This great cross, once established, divided the space into quadrants, probably to reflect the four quarters of the Mesoamerican universe. It also served as the basis of the great grid that eventually covered eight square miles—a plan so rigid that even the small river that crossed the area was rerouted to conform to it.

LEFT: The stone image of death, revealed near the Pyramid of the Sun, evokes the Mesoamerican underworld, represented at Teotihuacan by a series of natural caves. One, centered below the Pyramid of the Sun, may have helped planners determine the site of the great structure. OPPOSITE: The Pyramid of the Feathered Serpent, god of war, water, and agriculture, and thus the bringer of civilization, stood at the very center of Teotihuacan before it was buried beneath a later pyramid.

In "downtown" Teotihuacan, the Pyramid of the Sun, one of the largest structures ever built in ancient America, was raised over the main chamber system of the largest known cave at the site, and the main civic and religious complex of the city—an enormous square holding the ornate Pyramid of the Feathered Serpent—lay astride the all-important east-west axis.

Slowly the city grew outward, filling the great grid with stone-paved streets and alleys defining city blocks, each of which held one or more of Teotihuacan's apartment compounds. Some 2,000 of these compounds, grouped in distinct barrios, held the first true urban families of America. All compounds held chambers devoted not only to cooking, storage, sleeping, and craft activities, but also to shrines, the richest having elaborate murals dedicated to the Water Goddess, the Storm God, and others of the city's pantheon.

The work of archaeologist Linda Manzanilla and others who have meticulously analyzed the remains of Teotihuacan's compounds, indicates that each probably held two or more related families. Judging from their remains, the barrios included not only the Teotihuacanos themselves, but "foreigners" as well, including Zapotec from Oaxaca, Totonac from Veracruz, and Maya from

Dominated by the Temple of the Giant Jaguar, ruins of the Maya city of Tikal rise with the mist from the rain forest of northern Guatemala's Petén region. One of the largest cities of its times, Tikal prospered during the first millennium B.C.,

declined, then flourished throughout the Classic Period under a royal dynasty whose fortunes rose and fell during wars with Calakmul, Caracol, and other neighboring states. The city was ultimately abandoned around A.D. 900.

the distant lowlands of Yucatán. Occupations evident from remains of all parts of the city include potters and figurine makers, architects, obsidian workers, weavers, mural painters, sculptors, merchants, and traders.

Mass burials of sacrificial victims and probable royal tombs in the Pyramid of the Moon and Temple of Quetzalcoatl—both looted in antiquity—along with the sheer enormity of the public structures, indicate a succession of kings of extraordinary power. Other than that, we know almost nothing of Teotihuacan's leadership. The economic prosperity of the city lay in trade, particularly in a green-tinted obsidian of superb quality from the Cerro de las Navajas—Mountain of the Knives—some 35 miles to the north. This natural glass, so desired by ancient Mesoamericans, was traded from here to virtually every area of Mesoamerica.

At its greatest extent Teotihuacan covered some eight square miles—a space of harmony among its inhabitants, the land, the heavens, and time itself. In all its centuries of existence Teotihuacan never deviated from this grand cosmic scheme. Its enormous influence over the rest of Mesoamerica—at Monte Albán, at Xochicalco, and as far away as the Maya highlands of southern Guatemala—is just now coming to light.

NOWHERE WAS TEOTIHUACAN'S INFLUENCE more intense than in the Maya lowlands, the hearth of the Classic Period Maya, acknowledged by many as creators of one of the most intriguing civilizations in all of the ancient world.

Whether it happened through military conquest, actual colonization, trade, a religious movement, or some combination of all these factors, evidence of a Teotihuacan presence appears very early in the Early Classic architecture and ceramics in the Pacific slope area and in the Maya highlands of Guatemala, particularly at Kaminaljuyú, located in a sprawling suburb of Guatemala City. Somewhat later, Teotihuacan elements appear in the art and icons that mix with Maya motifs at sites scattered throughout the Maya lowlands, including Tikal, Copán, Palenque, and Uxmal. Indeed, the great Classic Period Maya regional capital of Copán

BELOW: *Ears of maize appear as divine beings in the murals of Cacaxtla, at Tlaxcala, Mexico, underscoring the immense importance of the food plant in the long history of Mesoamerican life and civilization. Developed from a wild grass by centuries of accident and experiment in the Mexican highlands, maize had transformed Mesoamerica into a farmer's world by 2000 B.C.*
OPPOSITE: *A modern Maya farmer harvests maize from his milpa, or cornfield, in the hill country of the northern Yucatán Peninsula.*

appears to have been virtually taken over by elite citizens of Teotihuacan, who, according to the Maya hieroglyphic texts, literally began the royal dynasty that helped to carry that city to greatness. One inscription at Copán, recently deciphered by David Stuart of Harvard University, states specifically that the first ruler of the Classic Period Copán dynasty, whose Maya name translates as Resplendent Quetzal-Macaw, appeared as a stranger at Copán—after having walked the trails from the "Place of the Reeds" (almost certainly a reference to Teotihuacan) for a total of 157 days!

As the example of Copán's Altar Q inscription demonstrates, our knowledge of the rise and fall of Maya civilization is much the greater because of the decipherment of the hieroglyphic writing. As we know mainly from the work of three remarkable scholars—Yuri V. Knorozov, Heinrich Berlin, and Tatiana Proskouriakoff—the system employed by the ancient Maya utilized different signs—pictures of geometrical forms, heads of living or supernatural beings, and other forms—to stand for either words or syllables. These could be carved in various combinations to construct texts of any kind needed—or numbers and dates in a system of date notation involving the concept of zero. Neither idea—writing or the calendar—originated with the Maya, but with their Preclassic forebears in the areas of Oaxaca, Veracruz, and the Pacific slope of southern Guatemala. It was the Maya scribe, however, that developed both the craft of writing and calendrical arithmetic to the highest degree known in all of ancient America.

By means of the hieroglyphs, ancient Maya scribes recorded the lives, political titles and offices, and the wars and exploits of Classic Period rulers and the nobility in general, along with virtually every other aspect of Maya court life and ritual. Skilled calligraphers painted scenes and texts on temple walls and on

vases used for the ritual drinking of chocolate. Sculptors transformed slabs of limestone into miraculous portraits and texts extolling the divine ancestry of the kings who held sway over the patchwork of Classic Period capitals and their surrounding states that stretched throughout the Yucatán Peninsula.

At Palenque, where Maya archaeology and epigraphy began in the 1820s, the hieroglyphs have revealed a distinguished dynasty, whose divine forebears, the three gods of the Palenque Triad, lived in the distant age before time began. The most prominent king, K'inich Janahb' Pakal, reigned from 615 to 683, when he was laid to rest in a giant carved sarcophagus in the Temple of the Inscriptions.

At Río Azul, Guatemala, not far from Tikal—which

LEFT: *Reconstructed from a pile of stones fallen from a building at the Maya ruins of Copán, Honduras, a maize god—or perhaps a king in the guise of the god—emerges from more than a millennium of darkness.*
OPPOSITE: *The Temple of the Inscriptions at Palenque, Chiapas, Mexico, served as the funerary memorial of that city's greatest ruler, K'inich Janahb' Pakal, who reigned from A.D. 615 to 683. His huge carved sarcophagus was discovered in 1952 at the end of a long rubble-choked stairway inside the pyramid.*

apparently held sway over it—the glyphs painted on tomb walls name the pyramids themselves as *witz*, or mountain, underscoring the fact that many if not most Maya and Mesoamerican cities were constructed as metaphors in stone. In this manner the pyramids stood for sacred mountains, their summit temples for caves, and their plazas and patios for bodies of water concealing Xibalbá, the Maya netherworld. Settlements and cities were often built directly over caves—as was the Pyramid of the Sun at Teotihuacan—which were believed to serve as entrances to the dreaded underworld. Ball courts complete the metaphor, for they served not only as places for the all-important ritual ball game, which itself symbolized the movement of celestial bodies, but also as interfaces between the real world and Xibalbá.

In more than a decade of intensive archaeology at Copán, in combination with textual data from the stelae and altars, Harvard archaeologist William Fash and his Honduran colleague Ricardo Agurcia Fasquelle have been able to glimpse the powerful dynasty founded by K'inich Yax K'uk' Mo', the stranger who appeared at

In one of the greatest surviving Maya murals of the Classic Period, the ruler Chaan-muan, holy lord of the state of Bonampak is clad in a jaguar-skin tunic. His spear in firm grasp, he dominates a scene of noble victors and bewildered

war captives. The event depicted covers a plastered wall in one of three painted rooms of a temple at Bonampak, Chiapas, Mexico, and took place in the late 8th century A.D., *following the capture of prisoners after a fierce battle, also depicted.*

139

Copán from Teotihuacan and acceded as king in the dry season of A.D. 426. The 15 rulers who followed carried the valley capital through some four centuries, a period that reached a high point of territorial expansion under the guidance of Oxlahun Ubaah K'awiil (nickname of convenience: 18 Rabbit), 13th ruler of the Copán state (reigned A.D. 695-738). Ironically, a monument at the rival capital of Quiriguá, some 30 miles to the north, records the capture and decapitation of this Copán ruler on May 3, 738—a humiliating event generally unacknowledged in the Copán texts.

The voice of the Maya hieroglyphic inscriptions, while doubtless tinged with elements of self-interest and the sort of exaggeration characteristic of most official writings, nonetheless provides us with a priceless glimpse of the concerns of the remarkable Maya—not to mention the dated history of the Classic Period.

Two key rulers of the royal dynasty of Copán, one dead, one alive, meet on the side of Altar Q, carved to mark the inauguration of Yax Pasah, Copán's last powerful king. In the scene, both are seated on their name hieroglyphs. Yax Pasah, right, receives the regalia of rulership from K'inich Yax K'uk' Mo', left, founder of the dynasty some 350 years before. The hieroglyphs between the two record the date— July 2, 763.

OPPOSITE: *Portrait in clay of an elaborately garbed Maya noble holding an unknown object is one of thousands known from Classic Period elite burials at such sites as Jaina, a small island off the west coast of the Yucatán Peninsula. Hand-modeled and hollow, such figurines also served as whistles.*

Teotihuacan began to decline shortly after A.D. 500, and by 750 had effectively vanished from the Mesoamerican scene. Burned structures at the city center suggest that an organized rebellion, perhaps a local move against a crippled elite bureaucracy, brought about the end. Other evidence hints at overpopulation in a land of limited bounty, maybe along with disease brought about by the sheer volume of sewage and accumulated waste among the apartment compounds.

Whatever the cause, the fall of Teotihuacan was felt throughout Mesoamerica. By 900 the Zapotec of Monte Albán had yielded to a Mixtec incursion from the mountains to the west and abandoned their Classic Period capital. In the Maya lowlands the raising of buildings and monuments ceased, seemingly along with the kind of elite rulership and institutions of royalty and religion that had proven so powerful through the Classic Period. After A.D. 900 virtually all the largest capitals, including Tikal, Calakmul, Copán, and Palenque, lay abandoned, victims, perhaps, of a series of fatal droughts which, archaeologist Richardson Gill persuasively argues, literally decimated the population. Others cite environmental degredation combined with overpopulation and the endemic warfare that was now carried out more in desperation for arable land than for prisoners to use as spectacles in the rites of public humiliation and sacrifice that took place on the pyramid steps.

The final six centuries of uninterrupted Mesoamerican cultural development culminated in a cosmopolitan world that emerged from the troubled end of the Classic Period. In the west the void created by the shattered power of Teotihuacan was filled by a succession of smaller, warring polities, from which Tula, the legendary Toltec capital in the dry lands north of the Basin of Mexico, emerged as a powerful warrior state near the beginning of the Postclassic Period. To the south of the Mexican Plateau the Mixtec welded a society ruled by hero kings such as Eight Deer-Jaguar Claw, whose genealogies were immortalized in brilliantly painted hieroglyphs that filled the screenfold books of deerskin. In addition, Mixtec craftspeople, working in copper and gold (metal and knowledge of its use had finally reached Mesoamerica from the region of Ecuador, probably via the Pacific coast of Mexico, in the 11th century) produced some of ancient America's finest jewelry for elite consumption.

BELOW: *Necklace of gold skulls linked by beads of shell and turquoise mark this as a Postclassic ornament. Knowledge of metalworking did not come to Mesoamerica until around A.D. 1000, probably from the Ecuador area by way of the Pacific coast. The style suggests its makers as the incredibly talented Mixtec crafts-*

people from the Oaxaca area of Mexico, who provided many luxury goods for Aztec royalty.
ABOVE: *Skilled mosaic work of turquoise and white-and-colored shell over hollowed wood mark this extraordinary 17-inch-wide double-headed serpent. For the Aztec, the serpent had many symbolic meanings. Among them, its undulating movement recalled the motion of water and thus fertility; the shedding of its skin symbolized renewal and transformation.*

In the Maya area, militant merchant states centered at Chichén Itzá and other cities continued the brisk commerce in salt, textiles, and other special commodities over the network of trails that laced the highlands and lowlands, while great cargo canoes hugged the coast as they went from port to port along the Gulf and Caribbean coasts from central Mexico to Honduras. In the course of such activity the Postclassic Maya became more cosmopolitan than their Classic Period ancestors, adapting an eclectic corpus of art motifs and styles from other parts of Mesoamerica. Throughout the period the shaman-kings and religious elite still wrote their histories and traditions, their astronomical notes and astrological almanacs in tall, narrow screenfold books of bark paper, and continued the noble necessity of maize farming that had helped start the whole system some 2,500 years earlier.

BACK IN THE HIGHLANDS to the west the people known to posterity as the Aztec entered the Basin of Mexico from an unknown home, Aztlan, or the White Land, at the beginning of the 13th century, led in their wanderings, tradition says, by an image of their god Huitzilopochtli. In 1325 they founded their city of Tenochtitlan

The famed "floating gardens" at Xochimilco, Mexico, are stationary rectangles of earth anchored by strategically placed trees. Laced by water-filled canals, such plots, built up by the excavation of marshy areas and the placement of rich soil so as to

form planting areas of extraordinary fertility, helped sustain the Aztec capital of Tenochtitlan. Mesoamerican farmers created terracing, ridged fields, and other agricultural innovations that sustained their cultures for more than 5,000 years. 145

on a marshy island in the middle of Lake Texcoco where, as a tribal prophecy had foretold, they encountered an eagle perched on a cactus, devouring a snake.

Unwelcome, and shunned by the civilized dwellers along the lakeshore, these opportunistic barbarians served others as vassals and mercenaries—and all the while, they learned. In the early 1400s the Aztec turned on their neighbors, conquered them, and began the consolidation of a tribute empire that by 1519 stretched as far as the border of Guatemala, some 500 miles to the southeast.

Aztec culture was an extravagant amalgam of all that had passed before in Mesoamerica. Aztec life was completely dominated by the viewpoint that saw humans as insignificant beings at the mercy of a universe filled with tribulations—ranging from fearsome earthquakes to solar eclipses and from drought to flood. Personal and national gods were profuse, and each was more likely than not to have multiple identities and functions in the hierarchy of supernaturals.

Wars were mainly ritual events that furnished sacrificial victims for the altars of Tenochtitlan, where the blood of the individuals so honored was necessary to keep the universe in motion. Aztec warfare also had another purpose—the subjugation of neighbors for tribute to the growing empire. As it had done continuously, the calendar ruled the Aztec cycles of religion and ritual. Special activities marked the Aztec months, and every 18,980 days—about 52 years—the New Fire ceremony, replete with the kindling of a new flame in the cut-open corpse of a sacrificial victim, marked the death of one life and, hopefully, nature's grant of another.

The Aztecs celebrated their final New Fire ceremony in 1507 of the Christian calendar. A few years later strangers began to appear in "wooden houses" that floated in the coastal waters. In 1519 the Aztec year One Reed began, a time that prophecies said would see the return of the Mesoamerican bearded culture hero Quetzalcoatl, or Feathered Serpent. Soon there came rumors from the east of four-legged monsters with humans growing from them, of invincible strangers with deadly weapons and a bearded leader. Was this Quetzalcoatl, fulfilling the prophecy? The meditative emperor Moctezuma II pondered the answer until it was too late, for on November 9 the conquistador Hernán Cortés and his army stood on a mountain pass to the east and there paused to contemplate Moctezuma's city with amazement.

The great city of Tenochtitlan and its twin, Tlatelolco, built on an island in the middle of Lake Texcoco, were linked by causeways and an aqueduct to the surrounding shores. A walled religious precinct marked the center of Tenochtitlan, dominated by the Great Temple, a double pyramid honoring Huitzilopochtli, patron god of the Aztec, and Tlaloc, god of storms and rain. Canals and streets subdivided the city into quadrants, each with its own temples, markets, and administrative buildings. From Tenochtitlan ten emperors ruled a tribute empire that extended over much of Mesoamerica and endured until its fall after the arrival of Hernán Cortés, who conquered the city in 1521.

CHAPTER 7

THE ANDES & BEYOND

HUNTERS, FISHERS, FARMERS, AND BUILDERS OF CIVILIZATION ACHIEVE DISTINCTION IN A VAST CONTINENT AND A VARIED ISLAND WORLD